D0960374

GENIUS FOODS

GENIUS FOODS

Become Smarter, Happier, and More Productive While Protecting Your Brain for Life

MAX LUGAVERE

WITH PAUL GREWAL, MD

HARPER WAVE

An Imprint of HarperCollins*Publishers*

This book contains advice and information relating to health care. It should be used to supplement rather than replace the advice of your doctor or another trained health professional. If you know or suspect you have a health problem, it is recommended that you seek your physician's advice before embarking on any medical program or treatment. All efforts have been made to assure the accuracy of the information contained in this book as of the date of publication. This publisher and the author disclaim liability for any medical outcomes that may occur as a result of applying the methods suggested in this book.

HarperCollins books may be purchased for educational, business, or sales promotional use. For information, please e-mail the Special Markets Department at SPsales@harpercollins.com.

FIRST EDITION

Designed by William Ruoto

Images on pages 168–169 courtesy of Mary Newport

Library of Congress Cataloging-in-Publication Data has been applied for.

ISBN 978-0-06-256285-2

18 19 20 21 22 LSC 10 9 8 7 6 5

This book is dedicated to the first genius I ever met: my mom.

CONTENTS

GENIUS FOODS

INTRODUCTION

Before you play two notes learn how to play one note—and don't play one note unless you've got a reason to play it.

—MARK HOLLIS

If you'd have told me a few years ago that I'd one day write a book about optimizing the brain, I would have been sure you had mistaken me for someone else.

After I switched my college major from premed to film and psychology, the idea of a career in health seemed unlikely. This was compounded by the fact that soon after I graduated, I became entrenched in what I considered a dream job: a journalist and presenter on TV and the Web. My focus was stories that I felt were underreported and could make a positive impact on the world. I was living in Los Angeles—a city I'd idolized as an MTV-watching teen growing up in New York—and had just ended a five-year stint hosting and producing content for a socially conscious TV network called Current. Life was great. And it was all about to change.

As much as I enjoyed the Hollywood life, I'd often find myself making trips back east to see my mom and two younger brothers. In 2010, on one of those trips home, my brothers and I noticed a subtle change in the way my mother, Kathy, walked. She was fifty-eight at the time and had always had a spirited way about her. But suddenly, it was as though she were wearing a space suit underwater—each stride and gesture looked like a purposeful, conscious decision. Though I know better now, back then I couldn't even make the connection between the way she moved and her brain's health.

She also began offhandedly complaining of mental "fogginess." This too was lost on me. No one in my family had ever had memory problems. In fact, my maternal grandmother lived to ninety-six and her memory was sharp until the end. But in my mom's case, it seemed as if her overall processing speed had slowed, like a Web browser with too many open tabs. We started to notice that when we would ask her to pass the salt at dinner, it would take her a few extra beats to register. While I chalked what I was seeing up to "normal aging," deep down I had the chilling suspicion that something wasn't right.

It wasn't until the summer of 2011 during a family trip to Miami that those suspicions were confirmed. My mom and dad had been divorced since I was eighteen, and this was one of the few times since then that my brothers and I were together with my parents under the same roof—seeking respite from the summer heat in my dad's apartment. One morning, my mother was standing at the breakfast bar. With the whole family present, she hesitated, and then announced that she had been having memory problems and had recently sought the help of a neurologist.

In an incredulous but playful tone, my father asked her, "Is that so? Well then, what year is it?"

She stared at us blankly for a moment, and then another.

My brothers and I chuckled and chimed in, breaking the uneasy silence. "Come on, how could you not know the year?"

She responded, "I don't know," and began to cry.

The memory is seared into my brain. My mom was at her most vulnerable, courageously trying to communicate her internal pain, defective but self-aware, frustrated and scared, and we were completely ignorant. It was the moment I learned one of life's hardest lessons: that nothing else means a thing when a loved one gets sick.

The flurry of medical visits, expert consultations, and tentative diagnoses that followed culminated at the tail end of a trip to the

Cleveland Clinic. My mom and I had just walked out of a renowned neurologist's office and I was trying to interpret the labels on the pill bottles clutched in my hand. They looked like hieroglyphics.

Staring at the labels, I silently mouthed out the drug names to myself in the parking lot of the hospital. *Ar-i-cept. Sin-e-met.* What were they for? Pill bottles in one hand, unlimited data plan in the other, I turned to the digital-age equivalent of a safety blanket: Google. In 0.42 seconds, the search engine returned results that would ultimately change my life.

Information on Aricept for Alzheimer's disease.

Alzheimer's disease? No one had said anything about Alzheimer's disease. I became anxious. Why hadn't the neurologist mentioned that? For a moment, the world around me ceased to exist but for the voice in my head.

Does my mom have Alzheimer's disease? Isn't that something only *old people* get?

How could she have it, and at this age?

Grandma is ninety-four and she's fine.

Why is mom acting so calm? Does she understand what this means? Do I?

How long does she have before . . . whatever comes next?

What *does* come next?

The neurologist had mentioned "Parkinson's Plus." *Plus* what? "Plus" had sounded like a bonus. Economy Plus means more legroom—usually a good thing. Pert Plus was shampoo plus conditioner, also a good thing. No. My mom was prescribed medicines for Parkinson's disease *plus* Alzheimer's disease. Her "bonus feature" was the symptoms of a bonus disease.

As I read about the pills I was still holding, repeating phrases stuck out to me.

"No disease-modifying ability."

"Limited efficacy."

"Like a Band-Aid."

Even the doctor had seemed resigned. (I later learned a cold joke circulated among med school students about neurology: "Neurologists don't treat disease, they admire it.")

That night I was sitting alone in our Holiday Inn suite, a couple of blocks from the hospital. My mom was in the other room, and I was at my computer, manically reading anything I could find on both Parkinson's and Alzheimer's disease, even though my mom's symptoms did not fit neatly into either diagnosis. Confused, uninformed, and feeling powerless, it was then that I experienced something I'd never felt before. My vision narrowed and darkened, and fear enveloped my consciousness. Even with my limited insight at the time I could tell what was happening. Heart pounding, hungry for air, a feeling of impending doom—I was having a panic attack. Whether it lasted minutes or hours I can't be sure, but even as the physical manifestations subsided the emotional dissonance remained.

I chewed on that sensation for days afterward. After I returned to LA and the initial storm cleared, I felt like I was left standing on a shattered landscape, surveying the path ahead without a map or compass. My mom began taking the chemical Band-Aids, but I felt continual unease. Surely the fact that we had no family history of dementia meant there had to be something environmental triggering her illness. What changed in our diets and lifestyles between my grandmother's generation and my mother's? *Was my mom somehow poisoned by the world around her?*

As these questions circled my head, I found little room to think about anything else, including my career. I felt like Neo from *The Matrix*, reluctantly conscripted by the white rabbit to save my mother. But how? There was no Morpheus to guide me.

I decided the first step was to pack up my West Coast life and move back to New York to be closer to my mom, so I did just that, and spent the following year reading everything I possibly could on both Alzheimer's and Parkinson's disease. Even in those early months, as I'd sit on her couch after dinner, face buried in research, I can recall watching my mom pick dishes up off the dining room table. Dirty plates in hand, she'd begin taking a few steps in the direction of her bedroom instead of the kitchen. I'd watch quietly, counting each second that would pass before she'd catch herself, as the knot that had formed in my stomach tied itself tighter. Every time, my fortitude in the search for answers was renewed.

One year turned into two, and two years turned into three, as my fixation on understanding what was happening to my mom consumed me. One day, it dawned on me that I had something that few others have: media credentials. I began to use my calling card as a journalist to reach out to leading scientists and clinicians around the globe, each of whom I've found to hold another clue in my scavenger hunt for truth. To date, I've read hundreds (if not thousands) of discipline-spanning scientific papers, and I've interviewed dozens of leading researchers and many of the most highly respected clinicians in the world. I've also had the opportunity to visit research labs at some of our most respected institutions—Harvard, Brown, and Sweden's Karolinska Institutet, to name a few.

What external environment allows our bodies and brains to thrive rather than malfunction? That became the basis of my investigation. What I've found has changed the way I think of our most delicate organ and defies the fatalistic view given to me by the vast majority of neurologists and scientific experts in the field. You will be surprised—perhaps even shocked—to learn that if you are one of the millions of people worldwide with a genetic predisposition to developing Alzheimer's disease (statistically, you have one-in-four odds of that being the case), you may respond *even better* to the principles proposed in this book. And, by following them, you will likely have more energy, better sleep, less brain fog, and a happier mood, *today*.

Through this journey, I've realized that medicine is a vast field with many silos. When it comes to knowing how best to care for something as complex as the human body, let alone the brain, you have to break apart those silos. Everything is related in unimaginable ways, and connecting the dots requires a certain level of creative thinking. You will learn about these many relationships in this book. For example, I'll share a method of fat burning so powerful some researchers have called it *biochemical liposuction*—and how it may be your brain's best weapon against decay. Or how certain foods and physical exercises actually make your brain cells work more efficiently.

While I'm dedicated to communicating the intricacies of nutrition to laypeople, I am also passionate about speaking directly to doctors, because surprisingly few are adequately trained in these topics. I've been invited to teach (as well as learn from!) medical students and neurology trainees at esteemed academic institutions such as Weill Cornell Medicine, and I have had the opportunity to lecture at the New York Academy of Sciences alongside many of the researchers cited in this book. I've helped create tools that are being used to teach physicians and other health-care providers around the world about the clinical practice of Alzheimer's prevention, and I've coauthored a chapter on the same topic in a textbook geared to neuropsychologists. I've even assisted with research at the Alzheimer's Prevention Clinic at Weill Cornell Medicine and NewYork-Presbyterian.

What follows is a result of this gargantuan and unending effort to understand not only what happened to my mother, but how to prevent it from happening to myself and others. My hope is that by reading about how to make your brain work better in the here and now, you will prevent your own decline and push your cognitive health span to its natural limits.

How to Use This Book

This book is a guide to attaining optimal brain function with the pleasant side effect of minimizing dementia risk—all according to the latest science.

Maybe you're looking to hit the reset button on your mental agility, to clear the *cache*, so to speak. Perhaps you're hoping to increase productivity and gain a leg up on your competition. Maybe you're one of the millions of people around the globe battling brain fog. Or depression. Or an inability to cope with stress. Perhaps you have a loved one suffering from dementia or cognitive decline and are fearful for them, or of succumbing to the same fate. No matter what led you to pick up *Genius Foods*, you're in the right place.

This book is an attempt to uncover the facts and propose new unifying principles to counter our collective modern malaise. You'll learn about the foods that have become casualties of the modern world—raw materials to build your best brain yet, replaced by the biological equivalent of cheap particleboard. Every chapter delves into the precise elements of optimal brain function—from your precious cell membranes, to your vascular system, to the health of your gut—all through the lens of what matters most: your brain. Each chapter is followed by a "Genius Food," containing many of the beneficial elements discussed in the surrounding text. These foods will serve as your weapons against cognitive mediocrity and decay—eat them, and eat them often. Later in the book, I'll detail the optimal Genius Lifestyle, culminating in the Genius Plan.

I've written this three-part book to be read cover to cover, but feel free to treat it as a reference and skip from chapter to chapter. And don't be afraid to take notes in the margins or highlight key points (this is often how I read!).

Throughout, you'll also find insights and "Doctor's Notes" highlighting my friend and colleague Dr. Paul Grewal's clinical and personal experience with many of the topics we'll cover. Dr. Paul has had his own challenges, having gone through medical school with what is now familiar to many in the Western world: obesity. Desperate to find a solution to his weight challenges, he ventured out to learn everything he could about nutrition and exercise—topics that are unfortunately all but ignored by med school curricula. The truths he discovered resulted in his shedding a dramatic one hundred pounds in less than a year, for good—and he'll be sharing these lessons on exercise and nutrition in the pages that follow.

Science is always unfinished business; it's a method of finding things out, not an infallible measure of truth. Throughout this book, we'll use our understanding of the best available evidence, while taking into account that not everything can be measured by a science experiment. Sometimes observation and clinical practice are the best evidence we have, and the ultimate determinant of health is how *you* respond to a given change. We take an evolutionary approach: we hold the position that the less time a food product or medicine or supplement has been around, the higher the burden of proof for it to be included in what we consider a healthful diet and lifestyle. We call this "Guilty until Proven Innocent" (see the section on polyunsaturated seed oils in chapter 2 as an example).

Personally, I started this journey from a blank slate, following the evidence wherever it took me. I've used my lack of preconceived notions to my advantage, to keep an objective distance from the subject and ensure that I've never missed the forest for the trees. Thus, you will see a linking of disciplines that may not be connected in other books of this genre, e.g., metabolism and heart health, heart health and brain health, brain health and how you actually *feel*. We believe that bridging these divides holds the keys to the cognitive kingdom.

Finally, we know that there are genetic differences between individuals, as well as differences in our health and fitness levels, that will determine things like carbohydrate tolerance and response to exercise. We've found the broadly applicable common denominators that will benefit everyone and have included sidebars with guidance on how to customize our recommendations to your own biology.

My hope is that when you finish reading *Genius Foods*, you'll understand your brain in a new way, as something able to be "tuned up" like a bicycle. You'll see food anew—as software, able to bring your brain back online and run your endlessly capable mind. You'll learn where to find the nutrients that can actually help you to remember things better and give you a greater sense of energy. You'll see that actually slowing the aging process (including cognitive aging) is just as much about the foods you omit from your diet as those you choose to consume, as well as *when* and *how* you consume them. I'll also share with you the food that may shave more than a decade off of the biological age of your brain.

I have to be honest—I'm so excited for you to begin this journey with me. Not only will you begin feeling your best within two weeks, you'll be fulfilling my hidden agenda—and perhaps my one true goal for you: to make use of the latest and best available evidence so that you might avoid what my mom and I have experienced. We deserve better brains—and the secret lies in our food.

The Genius Foods.

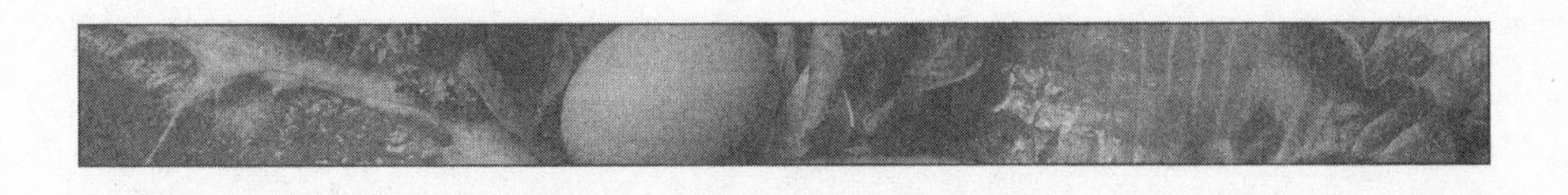

PART 1

You Are What You Eat

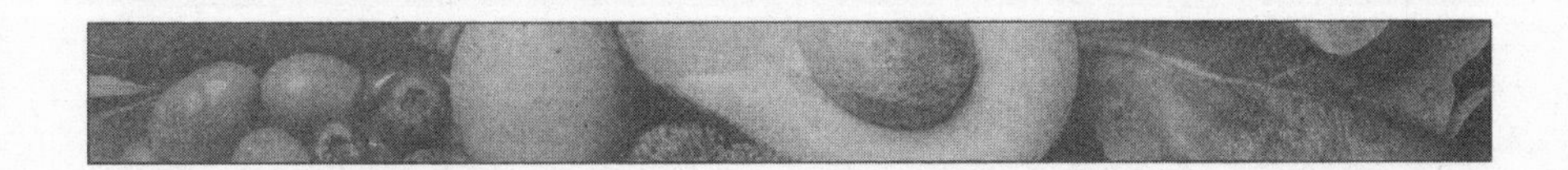

CHAPTER 1

THE INVISIBLE PROBLEM

Men ought to know that from the brain, and from the brain only, arise our pleasures, joys, laughter and jests, as well as our sorrows, pains, griefs and tears. Through it, in particular, we think, see, hear and distinguish the ugly from the beautiful, the bad from the good, the pleasant from the unpleasant. It is the same thing which makes us mad or delirious, inspires us with fear, brings sleeplessness and aimless anxieties. . . . In these ways I hold that the brain is the most powerful organ in the human body.

—HIPPOCRATES

Ready for the good news?

Nestled within your skull, mere inches from your eyes, are eighty-six billion of the most efficient transistors in the known universe. This neural network is *you*, running the operating system we know as life, and no computer yet conceived comes close to its awesome capabilities. Forged over millions and millions of years of life on Earth, your brain is capable of storing nearly *eight thousand iPhones' worth* of information. Everything you are, do, love, feel, care for, long for, and aspire to is enabled by an incredibly complex, invisible symphony of neurological processes. Elegant, seamless, and blisteringly fast: when scientists tried to simulate just one second of a human brain's abilities, it took supercomputers forty minutes to do so.

Now for the bad news: the modern world is like *The Hunger Games*, and your brain is an unwitting combatant, hunted mercilessly and relentlessly from all sides. The way we live today is undermining our incredible birthright, fighting our optimal cognitive performance, and putting us at risk for some seriously nasty afflictions.

Our industrially ravaged diets supply cheap and plentiful calories with poor nutrient content and toxic additives. Our careers shoehorn us into doing the same tasks over and over again, while our brains thrive with change and stimulation. We are saddled with stress, a lack of connection to nature, unnatural sleep patterns, and overexposure to news and tragedy, and our social networks have been replaced by The Social Network—all of which lead ultimately to premature aging and decay. We've created a world so far removed from the one in which our brains evolved that they are now struggling to survive.

These modern constructs drive us to compound the damage with our day-to-day actions. We convince ourselves that six hours in bed means we've gotten a full night's sleep. We consume junk food and energy drinks to stay awake, medicate to fall asleep, and come the weekend go overboard with escapism, all in a feeble attempt to grasp a momentary reprieve from our daily struggle. This causes a short circuit in our inhibitory control system—our brain's inner voice of reason—turning us into lab rats frantically searching for our next dopamine hit. The cycle perpetuates itself, over time reinforcing habits and driving changes that not only make us feel crappy, but can ultimately lead to cognitive decline.

Whether or not we are conscious of it, we are caught in the cross fire between warring factions. Food companies, operating under the "invisible hand" of the market, are driven by shareholders to deliver ever-increasing profits lest they risk irrelevance. As such, they market foods to us explicitly designed to create insatiable addiction. On the opposing front, our underfunded health-care system and scientific research apparatus are stuck playing catch-up, doling out advice and policy that however well intentioned is subject to innumerable biases—from innocuous errors of thought to outright corruption via industry-funded studies and scientific careers dependent on private-interest funding.

It's no wonder that even well-educated people are confused when it comes to nutrition. One day we're told to avoid butter, the next that we may as well drink it. On a Monday we hear that physical activity is the best way to lose weight, only to learn by Friday that its impact on our waistline is marginal compared to diet. We are told over and over again that whole grains are the key to a healthy heart, but is heart disease *really* caused by a deficiency of morning oatmeal? Blogs and traditional news media alike attempt to cover new science, but their coverage (and sensational headlines) often seems more intent on driving hits to their websites than informing the public.

Our physicians, nutritionists, and even the government all have their say, and yet they are consciously and subconsciously influenced by powers beyond the naked eye. How can you possibly know who and what to trust when so much is at stake?

My Investigation

In the early months following my mother's diagnosis, I did what any good son would do: I accompanied her to doctors' appointments, journal full of questions in hand, desperate to attain even a sliver of clarity to ease our worrying minds. When we couldn't find answers in one city, we flew to the next. From New York City to Cleveland to Baltimore. Though we were fortunate enough to visit some of the highest-ranking neurology departments in the United States, we were met every time with what I've come to call "diagnose and adios": after a battery of physical and cognitive tests we were sent on our way, often with a prescription for some new biochemical Band-Aid and little else. After each appointment, I became more and more obsessed with finding a better approach. I lost sleep to countless late-night hours of research, wanting to learn everything I possibly could about the mechanisms underlying the nebulous illness that was robbing my mom of her brainpower.

Because she was seemingly in her prime when her symptoms first struck, I wasn't able to blame old age. A youthful, fashionable, and charismatic woman in her fifties, my mom was not—and still is not—the picture of a person succumbing to the ravages of aging. We had no prior family history of any kind of neurodegenerative disease, so it seemed her genes could not be solely responsible. There had to be some external trigger, and my hunch was that it had something to do with her diet.

Following that hunch led me to spend the better part of the past decade exploring the role that food (and lifestyle factors like exercise, sleep, and stress) play in brain function. I discovered that a few vanguard clinicians have focused on the connection between brain health and metabolism—how the body creates energy from essential ingredients like food and oxygen. Even though my mom had never been diabetic, I dove into the research on type 2 diabetes and hormones like insulin and leptin, the little-known signal that controls the body's metabolic master switch. I became interested in the latest research on diet and cardiovascular health, which I hoped would speak to the maintenance of the network of tiny blood vessels that supply oxygen and other nutrients to the brain. I learned how the ancient bacteria that populate our intestines serve as silent guardians to our brains, and how our modern diets are literally starving them to death.

As I uncovered more and more about how food plays into our risk for diseases like Alzheimer's, I couldn't help but integrate each new finding into my own life. Almost immediately, I noticed that *my* energy levels began to increase, and they felt more consistent throughout the day. My thoughts seemed to flow more effortlessly, and I found myself in a better mood more often. I also noticed that I was more easily able to direct my focus and attention and tune out distractions. And, though it wasn't my initial goal, I even managed to lose stubborn fat and get in the best shape of my life—a welcome bonus! Even though my research was initially motivated by my mom, *I* became hooked on my new brain-healthy diet.

I had inadvertently stumbled upon a hidden insight: that the same foods that will help shield our brains against dementia and aging will also make them work *better* in the here and now.[1] By investing in our future selves, we can improve our lives *today*.

Reclaim Your Cognitive Birthright

For as long as modern medicine had existed, doctors believed that the anatomy of the brain was fixed at maturity. The potential to change—whether for a person born with a learning disability, a victim of brain injury, a dementia sufferer, or simply someone looking to improve how their brain worked—was considered an impossibility. Your cognitive life, according to science, would play out like this: your brain, the organ responsible for consciousness, would undergo a fierce period of growth and organization up to age twenty-five—the peak state of your mental hardware—only to begin a long, gradual decline until the end of life. This was, of course, assuming that you didn't do anything to accelerate that process along the way (hello, college).

Then, in the mid-nineties, a discovery was made that forever changed the way scientists and doctors viewed the brain: it was found that new brain cells could be generated throughout the life of the adult human. This was certainly welcome news to a species heir to the flagship product of Darwinian evolution: the human brain. Up until that point, the creation of new brain cells—called *neurogenesis*—was thought to occur only during development.[2] In one fell swoop, the days of "neurological nihilism," a term coined by neuroscientist Norman Doidge, were over. The concept of lifelong neuroplasticity—the ability of the brain to change up until death—was born, and with it a unique opportunity to mine this landmark discovery for greater health and performance.

Flash forward just a couple of decades to today and you could almost develop whiplash from the progress being made toward the understanding of our brains—both how we can protect them and how we can enhance them. Take the developments in the field of Alzheimer's disease research. Alzheimer's is a devastating neurodegenerative condition affecting more than five million people in the United States (with numbers expected to triple in the coming years); it is only recently that diet was thought to have any impact on the disease at all. In fact, though the disease was first described in 1906 by German physician Alois Alzheimer, 90 percent of what we know about the condition has been discovered in just the last fifteen years.

GIVE DEMENTIA THE FINGER

I had the privilege of visiting Miia Kivipelto, a neurobiologist at Stockholm's Karolinska Institutet and one of the foremost researchers exploring the effects of diet and lifestyle on the brain. She leads the groundbreaking FINGER trial, or Finnish Geriatric Intervention Study to Prevent Cognitive Impairment and Disability, the world's first ongoing, large-scale, long-term randomized control trial to measure the impact that our dietary and lifestyle choices have on our cognitive health.

The trial involves over 1,200 at-risk older adults, half of whom are enrolled in nutritional counseling and exercise programs, as well as social support to reduce psychosocial risk factors for cognitive decline such as loneliness, depression, and stress. The other half—the control group—receives standard care.

After the first two years, initial findings were published revealing striking results. The overall cognitive function of those in the intervention group increased by 25 percent compared to controls, and their *executive function* improved by 83 percent. Executive function is critically important to many aspects of a healthy life, playing a key role in planning, decision making, and even social interaction. (If your executive function isn't working up to snuff, you might complain that you are unable to think clearly or "get stuff done.") And the volunteers' *processing speed* improved by a staggering 150 percent. Processing speed is the rate at which one takes in and reacts to new information, and it typically declines with age.

The success of this trial highlights the power that a full lifestyle "makeover" can have on improving the way your brain works, even in old age, and provides the best evidence to date that cognitive decline does not have to be an inevitable part of aging.

As a result of this shift in our understanding of the brain, institutions such as the Center for Nutrition, Learning, and Memory at the University of Illinois Urbana-Champaign have sprung up, dedicated to filling in the gaps of our collective neuro-knowledge. Other emerging specialties have followed suit, eager to explore the links between our environments (including diet) and various aspects of our brain function. Take Deakin University's Food and Mood Centre, which exists solely to study the link between diet and mood disorders. In 2017, the center revealed how even *major depression* might be treated with food. I'll detail these findings, and the exact foods that can boost your mood, in the chapters to come.

Still, many remain in the dark about this vast and rapidly growing body of research. A study performed by AARP found that while over 90 percent of Americans believe brain health to be very important, few know how to maintain or improve it. Even our own well-meaning physicians, to whom we turn when we're scared and confused, are seemingly behind the times. The *Journal of the American Medical Association* itself reported that it takes seventeen years on average for scientific discoveries to be put into day-to-day clinical practice.[3] And so, we move through the motions as the old narrative continues—but it doesn't have to be this way.

A Genetic Master Controller—You!

Without imperfection, neither you nor I would exist.

—STEPHEN HAWKING

Mistakes is the word you're too embarrassed to use. You ought not to be. You're a product of a trillion of them. Evolution forged the entirety of sentient life on this planet using only one tool: the mistake.

—ROBERT FORD (PLAYED BY ANTHONY HOPKINS), *WESTWORLD*, HBO

Our genes were once considered our biological playbook—the code that ran our lives, including how our brains functioned. Understanding this code was the goal of the Human Genome Project, completed in 2002, with the hope that by the end, the secrets to curing human disease (including cancer and genetic diseases) would be splayed out in front of us. Though the project was a remarkable scientific achievement, the results were disenchanting.

It turns out that what distinguishes one person from the next is actually quite insignificant from a genetic standpoint, accounting for less than 1 percent of total genetic variation. So then why do some people live well into their nineties and beyond, maintaining robust brains and bodies, while others do not? Questions like this have continued to perplex scientists in the wake of the project, and have given rise to the idea that there has to be some other factor, or factors, to account for the wide range of differences in health and aging displayed by the global human population.

Enter *epigenetics*, the phoenix to rise from the project's ashes. If our genes are akin to the keys on a grand piano with twenty-three thousand notes, we now understand that our choices are able to influence the song that is played. This is because while our choices can't change our hard-coded genetics, they can impact the layer of chemicals that sits atop our DNA, telling it what to do. This layer is called the *epigenome*, derived from the Greek word *epi*, which means "*above*." Our epigenome affects not only our chances of developing whatever disease we're most at risk for, but also the moment-to-moment expression of our genes, which respond dynamically to the countless inputs we give them. (Perhaps even more shrouded in mystery is the sheet music, the order and sequence and frequency of activation of every gene in the development of a given organism—but that is for another book!)

While a treatise on epigenetics could span volumes, this book will zero in on one of the marquee maestros to play on our genetic keyboard: diet. Will your genetic conductor be a Leonard Bernstein, or a fifth-grade student pounding the ivories for the first time? It may depend largely on your dietary choices. What you eat will determine whether you'll be able to modulate inflammation, "train" a prizewinning immune system, and produce powerful brain-boosting compounds—all with the help of a few underappreciated nutrients (and lifestyle techniques) that have become seemingly lost to the modern world.

As you proceed, remember: nobody's a perfect specimen. I'm certainly not, and neither is Dr. Paul (though he'd argue otherwise). When it comes to genes, everyone has traits that, when thrust up against the modern world, increase their risk for cardiovascular disease, cancer, and, yes, dementia. In the past, these differences may have driven the evolution of our species, serving as advantages in our mysterious ancestral world. Now, these differences are why any person that makes it to age forty has an 80 percent chance of dying from one of these maladies. But, there's good news: if there's anything the past few years have shown us, it's that genes are not destiny—they merely predict what the Standard American Diet will do to you. This book will position you to be in the 20 percent as we address how to keep your brain and vascular system healthy (and even check some boxes for cancer prevention and weight loss while we're at it).

In the next few chapters, I'll describe an evidence-based antidote to the brain-shrinking Standard American Diet and lifestyle, replete with nutrients to fuel your ravenous brain and physical and mental techniques to take back the robustness that is your evolutionary destiny. Your primary opponents in the battle for your cognitive birthright are inflammation, overfeeding, nutrient deficiencies, toxic exposure, chronic stress, physical stagnation, and sleep loss. (If this sounds like a lot, don't worry—they overlap, and tending to one often makes it easier to improve on the others.)

Here's a brief overview of each of these "bad guys."

Inflammation

In a perfect world, inflammation is simply the ability of our immune systems to "spot clean" cuts, wounds, and bruises and to prevent the occasional bacterial tourist from becoming a full-blown infection. Today, our immune systems have become chronically activated in response to our diets and lifestyles. This has been recognized in the past several years as playing a pivotal role in driving or initiating many of the chronic, degenerative diseases plaguing modern society. Widespread inflammation can eventually damage your DNA, promote insulin resistance (the underlying mechanism that drives type 2 diabetes), and cause weight gain. This may be why systemic inflammation correlates significantly with a larger waistline.[4] In the coming chapters, we will definitively link these same factors to brain disease, brain fog, and depression as well.

Overfeeding

We haven't always been able to summon our food with a few swipes on a smartphone. By solving our species' food scarcity problem during the Agricultural Revolution, we've created a new one: overfeeding. For the first time in history, there are more *overweight* than *underweight* humans walking the Earth.[5] With our bodies constantly in a "fed" state, an ancient balance has been lost, one that has set us up for low brain energy, accelerated aging, and decay. Part of this has to do with the fact that many foods today are specifically designed to push our brains to an artificial "bliss point" beyond which self-control becomes futile (we'll explore this in chapter 3).

Nutrient Deficiency

In *Vanilla Sky* (one of my favorite films), writer/director Cameron Crowe wrote, "Every passing minute is a chance to turn it all around." This is particularly true of our bodies' ability to repair against the damages incurred due to aging, but *only* when we feed them the right ingredients. With 90 percent of Americans now falling short in obtaining adequate amounts of at least one vitamin or mineral, we have set the stage for accelerated aging and decline.[6]

Toxic Exposure

Our food supply has become awash in "food-like" products. These products directly contribute to the three factors mentioned above: they are stripped of nutrients during the production process, they promote their own overconsumption, and they drive inflammation. Most insidious, however, may be the "bonus" toxic additives—the syrups, industrial oils, and emulsifiers that directly and indirectly contribute to an activated immune system, driving anxiety, depression, suboptimal cognitive performance, and long-term risk for disease.

Chronic Stress

Chronic psychological stress is a major problem in the Western world. Like inflammation, the body's stress response was designed by evolution to keep us safe, but it has been hijacked by the modern world. While chronic stress is directly toxic to our brain function (covered in chapter 9), it also sends us reaching for unhealthy foods, thus compounding the damage done.

Physical Stagnation

Our bodies are designed to move, and ignoring that fact causes our brains to suffer. The evidence on exercise is mounting to an impressive degree, validating it not only as a method of boosting our long-term brain health (enabling us to fend off diseases once thought unpreventable), but as a means of enhancing the way we think and learn.

Similarly, we've evolved with another type of exercise: thermal exercise. We are great at changing our environment to suit our comfort levels, but the relative lack of variation in temperature we experience on a daily basis may undermine our peak brainpower and resistance to disease.

Sleep Loss

Last but not least, good-quality sleep is a precondition for optimal brain function and health. It gives you the *ability* to make dietary and lifestyle changes by making sure your hormones are working for you, not against you. And it purifies your brain and backs up your memories. *Costco*-size gains for *dollar store* effort, and yet our collective sleep debt is rising.

As I've mentioned, any one of these villains has the power to wreak cognitive destruction, and they have formed an unholy alliance to do so. But, if you allow this book to be your bow and arrow, sword and spear, you may just stand a chance.

In the coming chapters, we'll lay out a road map to circumvent the shortcomings of our discordant, high-stress lifestyles as we marry evolutionary principles with the latest clinical research. We'll use diet to reset your brain to its "factory settings," leaving you feeling and performing your very best. And we'll even venture into the new and exciting science surrounding the microbiome—the collective of bacteria that live within us, working the knobs and levers of our health, mood, and performance in astounding ways. They provide a new lens from which to assess our every choice.

Up next, as you begin to reclaim your cognitive legacy, you're going to learn about the nutrient that your brain is desperate for. May the odds be ever in your favor.

GENIUS FOOD #1

EXTRA-VIRGIN OLIVE OIL

Place some extra-virgin olive oil (EVOO) in a spoon, and then slowly slurp it up like you're eating soup and being particularly rude about it. (Yes, I'm telling you to drink oil, but you'll see why in a second.) You should in short order notice a spicy feeling in the back of your throat: that's a compound called *oleocanthal.* Oleocanthal is a type of phenol—plant compounds that powerfully stimulate our bodies' own repair mechanisms when we consume them (phenols are usually found linked together in the form of polyphenols). Oleocanthal possesses anti-inflammatory effects so powerful that it is comparable to taking a small dose of ibuprofen, a nonsteroidal anti-inflammatory drug, but without any of the potential side effects.[1] Inflammation, as you'll learn, can strongly negate neuroplasticity (the ability of the brain to change throughout life) and even produce feelings of depression, as research is now beginning to show.

Extra-virgin olive oil is a staple food in the Mediterranean diet, and people who consume these kinds of diets display lower incidence of Alzheimer's disease. Oleocanthal may play a role here as well, having demonstrated the potential to help the brain clear itself of the *amyloid* plaque, the sticky protein that aggregates to toxic levels in Alzheimer's disease.[2] It does this by increasing the activity of enzymes that degrade the plaque. It has been shown in large, long-term trials to protect the brain against decline (and even improve cognitive function) when consumed at volumes of up to a liter per week.[3] And if protecting your brain wasn't enough, EVOO has been shown to block an enzyme in fatty tissue called *fatty acid synthase*, which creates fat out of excess dietary carbohydrates.[4]

Aside from oleocanthal, EVOO is also a rich source of monounsaturated fat, which is a healthy fat that helps maintain the health of your blood vessels and your liver, and can even help you lose weight. One tablespoon also contains 10 percent of the recommended intake of vitamin E per day. Vitamin E is an antioxidant that protects fatty structures in your body—such as your brain—from the wear and tear of aging.

Nicholas Coleman, one of the world's few *oleologists* specializing in the cultivation of ultra-premium extra-virgin olive oils, had a few tips to share with me about finding the right olive oil. For one, color has no bearing on the quality of the oil. The single best way to assess an oil is to taste it. Good extra-virgin olive oils should taste grassy, never greasy. Because oleocanthal is responsible for virgin oil's peppery taste, it can in fact be used as a measure of how much oleocanthal is present in the oil. Stronger oils can be so spicy that you may find yourself coughing from the heat—which is actually a classification of oil quality! Next time you find yourself consuming "three-cough" oil, you'll know you've found a keeper and your brain will thank you for it.

How to use: Extra-virgin olive oil should be the main oil in your diet, to be used liberally on salads, eggs, and as a sauce. Ensure that the oil is kept in a bottle that shields it from light (dark glass or tin is fine) and store in a cool, dry place.

CHAPTER 2

FANTASTIC FATS AND OMINOUS OILS

My recollections from childhood in the late eighties and nineties have a few standout milestones: singing the lyrics to the *Teenage Mutant Ninja Turtles* theme song on repeat (turtle power!), my first *Ghostbusters* Halloween outfit, and waking up at an ungodly hour on Saturday mornings to watch one of the first great serials of the modern television renaissance: *X-Men: The Animated Series*.

My memory of my family's dietary pattern is a little less vivid. Meals in my house were often prepared by my mom, who was as health-conscious as any busy woman with three little boys (and a fourth, if you count my dad) could be. She watched the *Nightly News*, read the *New York Times* and various magazines, and was generally hip to the mainstream health advice of the time. There was no social media, but TV and magazines did a pretty good job of relaying the latest discoveries and government recommendations. This was how many, including my mom, got their ideas about nutrition.

The main cooking oils in my home were canola oil and corn oil, because they were cholesterol-free and had no saturated fat. Many nights, dinner consisted of some kind of wheat-based noodle or spaghetti, tossed in margarine—the purported health alternative to "artery-clogging" butter. It was a dish that would win the heart of any dietician of the early nineties.

Unfortunately, the totality of my mother's—and, in all likelihood, your family's—concept of "diet" at the time was the end result of misguided nutrition science, biased government policy, and business doing what it does best—cost-cutting, lobbying, and marketing. And it was all total bullshit.

It began in the fifties, when Americans were hungry for a solution to an increasingly urgent public health problem: heart disease. My mom, born in 1952, grew up in the midst of what must have seemed like a terrible national epidemic. Heart disease was believed to be an "inevitable accompaniment of aging" and something that physicians could do little about.[1] In *The Big Fat Surprise*, food journalist Nina Teicholz recounts the furor: "A sudden tightening of the chest would strike men in their prime on the golf course or at the office, and doctors didn't know why. The disease had appeared seemingly out of nowhere and had grown quickly to become the nation's leading cause of death." That is until one outspoken scientist emerged from the dark halls of academia with a candle.

His name was Ancel Keys, a pathologist at the University of Minnesota. Even though Keys was not a medical doctor, he had attained a bit of nutritional "street cred" during World War II when he created the K ration, a system of boxed meals delivered to soldiers on the battlefield. After the war, Keys was enlisted by the Minnesota Health Department to contemplate the country's sudden cardiovascular quandary. Keys's hypothesis was that dietary fat was at the center of the epidemic, and to illustrate this, he drew up a graph from national data that depicted a perfect correlation between total calories consumed from fat and death rates from heart disease. Six countries were included.

Ancel Keys is often credited for having set off the domino effect that sculpted nutrition policy for the next sixty years, but his argument was built on data that was biased and ultimately misconstrued. His graph highlighted a correlation between two variables that had been cherry-picked amid the endless sea of variables that one encounters when studying things like diet across the population scale. But correlations cannot prove causation; they can only show relationships that are used to guide further study. In this case, however, the causal presumption was made, turning Keys into a national hero and landing him on the cover of *Time* magazine in 1961.

As it was taking a foothold in the national conversation, there was a growing chorus of voices within the scientific community who saw through Keys's work. Many thought that the validity of Keys's correlation itself was questionable: he omitted data that was available from *sixteen* other countries that, if included, would have shown no such correlation. For example, there was no epidemic of heart disease in France, a country whose citizens love their cheese and butter—the so-called French paradox. Others doubted that there was any link between fat consumption and heart disease at all.

John Yudkin, founding professor of the Department of Nutrition at Queen Elizabeth College in London, was one of Keys's vocal dissenters. As early as 1964, Yudkin thought sugar was the culprit, not fat. He wrote, "In the wealthier countries, there is evidence that sugar and sugar-containing foods contribute to several diseases, including obesity, dental caries [cavities], [type 2] diabetes mellitus and myocardial infarction [heart attack]." Reanalysis of Keys's data many years later confirmed that sugar intake had always correlated more strongly with heart disease risk than any other nutrient. After all, refined sugar, up until the 1850s, had been a rare treat for most people—a luxury, often given as a gift—but we'd been consuming butter for millennia.

Another researcher, Pete Ahrens, expressed similar befuddlement. His own research suggested that it was the carbohydrates found in cereals, grains, flour, and sugar that might be contributing directly to obesity and heart disease. (Research decades later would link these very factors to brain disease as well.) But Yudkin, Ahrens, and their colleagues failed in their attempts to speak above the "charismatic and combative" Keys, who also happened to have a powerful secret ally.[2]

In 1967, a review of the dietary causes of heart disease was published in the prestigious *New England Journal of Medicine* (*NEJM*). It was a no-holds-barred takedown, singling out dietary fat (and cholesterol) as the chief cause of heart disease. The role of sugar was minimized in the widely read paper, taking the wind out of the sails of any who would try to debate Keys. But reviews of this kind (and scientific research in general) are meant to be objective, not swayed by the influence of money. While researchers often rely on outside funding, in such cases the sources of their funding must be disclosed to alert their peers of any potential bias. Unfortunately, this was not the case with that *NEJM* article. The scientists behind it were each paid the equivalent of $50,000 in today's money by a trade organization called the Sugar Research Foundation (now known as the Sugar Association)—a fact that was *not* disclosed in the original paper. Worse yet, the foundation even influenced the selection of studies reviewed by the scientists. "They were able to derail the discussion about sugar for decades," said Stanton Glantz, a professor of medicine at University of California, San Francisco, in an interview with the *New York Times*. Dr. Glantz published these findings in the *Journal of the American Medical Association* in 2016.[3] (If you'd like to think such nefarious tactics are behind us, think again. The sugar industry continues to muddy the scientific waters, funding research that seems to conveniently conclude that claims against sugar are overhyped.)[4]

ENTER: FRANKENFOOD

To what degree can food be manipulated before we can no longer call it food? For many years, products that didn't adhere to the strict definitions of basic foods had to be labeled as "imitation." But a label bearing the *i*-word spelled marketing doom for products, so the food industry lobbied to have this imposition deregulated. In 1973 they got what they wanted. In *In Defense of Food*, journalist Michael Pollan writes:

> The regulatory door was thrown open to all manner of faked low-fat products: Fats in things like sour cream and yogurt could now be replaced with hydrogenated oils or guar gum or carrageenan, bacon bits could be replaced with soy protein, the cream in "whipped cream" and "coffee creamer" could be replaced with corn starch, and the yolks of liquefied eggs could be replaced with, well, whatever the food scientists could dream up, because the sky was now the limit. As long as the new fake foods were engineered to be nutritionally equivalent to the real article, they could no longer be considered fake.

Suddenly, the Frankenfood floodgates were open and the food supply was awash in fake products. It was like the breaking of the portal to hell in the 1987 movie *The Gate*, but instead of an onslaught of ghoulish creatures, they were the processed doppelgängers of real foods, now accompanied by a low-fat or fat-free halo.

One of the more absurd products of the flock came out in the late nineties: potato chips that had been formulated with the molecule olestra. It was a dream come true—a lab-created fat substitute that miraculously slid through your digestive tract unabsorbed. The only downside? The cramps, bloating, and "anal leakage" that created the equivalent of the Exxon Valdez oil spill in unsuspecting underpants everywhere.

How do you avoid Frankenfood when even today, a trip to the modern supermarket is the equivalent of skipping across a field of land mines? Shop around the perimeter of your supermarket, which is usually where the perishable, fresh food can be found; it's the aisles where Frankenfood usually lies in wait. And stick to the Genius Foods as well as the extended shopping list in the Genius Plan in chapter 11. (You can find a comprehensive guide to surviving the modern supermarket at http://maxl.ug/supermarkets.)

Eventually, Keys published the *Seven Countries Study*, a landmark research achievement albeit with similar flaws to his earlier work. In it, Keys shifted focus from total fat consumption to saturated fat. Saturated fat is solid at room temperature and is found in foods like beef, pork, and dairy. As anyone who's ever poured grease down a drain knows, this type of fat can clog pipes—and to an America at the dawn of nutrition science, it made perfect sense that this would happen in the body as well (spoiler alert: it doesn't).

Newly focused on these "artery-clogging" fats, Keys was able to influence a little-known organization (at the time) called the American Heart Association. With an investment from a massive manufacturing conglomerate called Procter & Gamble that produced, among other things, *polyunsaturated* vegetable oils (which are highly processed and, unlike saturated fats, liquid at room temperature), the organization finally had the ability to become a national powerhouse. It bought up TV and magazine ads alerting Americans to the boogeyman hiding in their butter. When the US government adopted the idea in 1977, the "low-fat" myth became gospel.

In an instant, Americans became the target of manufacturers seizing the opportunity to churn out "healthy" low-fat, high-sugar foods and polyunsaturated fat-based spreads ("cholesterol-free!"). Chemical- and heat-extracted oils like canola and corn oil were promoted to health food status, while naturally occurring fats from whole foods—even avocados—were shunned. Overnight, margarine—a rich source of a synthetic fat called *trans fat*—became "heart-healthy buttery spread."

Between industry shortcuts, scientific hubris, and governmental ineptitude, we took real, natural foods and warped them into a chemical minefield of "nutrients." The first victim of this fatty fiasco? Our brains, which are made almost entirely of fat. Sixty percent of the delicate, damage-prone human brain is composed of fatty acids, and—as we'll see in the following pages—the kinds of fats you consume dictate both the moment-to-moment quality of your brain's function and its predilection to disease.

Fats take a marquee role in every aspect of your life—from your decision-making processes, to your ability to lose weight, to your risk for diseases like cancer, and even the rate at which you age. By the end of this chapter, you'll be able to choose the fat-containing foods that optimize not only your cognitive performance, executive function, mood, and long-term brain health, but your overall health as well. If you take anything away from this section, let it be that it's not the amount of fat you consume; it's the type.

Polyunsaturated Fats: The Double-Edged Sword

Polyunsaturated fats are a type of dietary fat that's ubiquitous in our brains and bodies. The most well-known polyunsaturated fats are the omega-3 and omega-6 fats, which are considered essential because our bodies need them, and we can't produce them on our own. We therefore have to get these fats from food.

Two of the most important omega-3 fats are eicosapentaenoic acid (EPA) and docosahexaenoic acid (DHA). These are the "good" fats found in fish such as wild salmon, mackerel, and sardines, in krill, and in certain algae. They're also found in smaller amounts in grass-fed beef and pasture-raised eggs. While EPA is a whole-body anti-inflammatory agent, DHA is the most important and abundant structural component of healthy brain cells. Another form of omega-3 found in plants is called *alpha-linolenic acid* (ALA). ALA needs to be converted to EPA and DHA to be utilized by your cells, but the body's ability to do this is highly limited and varies in efficacy from person to person (I'll come back to this).

On the other side of the polyunsaturated coin, we have the omega-6 fatty acids. These are also essential to a healthy brain, but the American diet now includes far too many of them in the form of linoleic acid. These omega-6 fats went from appearing in our diet as oils delivered in trace quantities by whole foods to becoming major caloric contributors to the American diet in just a few short decades. They are the predominant type of fatty acid found in the grain and seed oils that we now consume in excess: safflower, sunflower, canola, corn, and soybean oils.

Night of the Lipid Dead

As coveted as polyunsaturated fats are by the brain, they are delicate and highly vulnerable to a process called *oxidation*. Oxidation occurs when oxygen (you may have heard of it) reacts chemically with certain molecules to create a new, damaged "zombie" molecule that has a super-reactive extra electron, called a *free radical*. How reactive is "super-reactive"? Let's just say these radicals make the White Walkers from *Game of Thrones* look like a caravan of pacifist hippies.

That extra electron can then react with another nearby molecule, transforming it into a second free radical and setting off a never-ending chain reaction that leaves utter mayhem in its wake. It's the biochemical equivalent of the zombie apocalypse, one molecule biting and infecting the one next to it, spawning a horde of the undead. Pioneering Austrian organic biochemist Gerhard Spiteller, who had done much of the eye-opening research on the dangers of oxidized polyunsaturated fats, put it this way:

> Radicals are typically four orders of magnitude (10,000x) more reactive than non-radical molecules. Their action is not under genetic control, they attack nearly all biological molecules, destroying lipids, proteins, nucleic acids [DNA], hormones and enzymes until the radicals are quenched by scavenger molecules.

This is a form of chemical damage that all organic matter is subject to, like rust on iron (iron is actually a catalyst for this same process in the human body, and may partly explain why men get more and earlier heart disease than women: they have more red blood cells, and more iron in circulation) or a sliced apple that has turned brown. Leave an apple slice on the counter for a few minutes and you can appreciate just how quickly these chemical reactions take place. In the body, excessive oxidation equals inflammation and damage to cellular structures and DNA. It's also thought to be one of the primary mechanisms of aging.

The battle against oxidation is a constant game of tug-of-war for all living creatures. Our own bodies, when healthy, have built-in antioxidant defense capabilities, and ideally, we churn out antioxidants—the aforementioned scavenger molecules—as fast or faster than free radicals can be created. (Many of the Genius Foods are beneficial in part because they increase your body's production of its own scavenger molecules.) Chronic inflammation or diseases like type 2 diabetes impair our ability to fight the accumulation of oxidative stress, and this is compounded when we absorb excess pro-oxidants from our food. It only takes a small amount of oxidative stress to set off a nuclear chain reaction of biochemical destruction, and the balance is a delicate one.

This places the brain in a unique and precarious situation. Accounting for 20 to 25 percent of your body's oxygen metabolism, constructed in large part by these delicate polyunsaturated fats, and squeezed into a container the size of a grapefruit, it couldn't be a larger magnet for oxidation. When oxidative stress overwhelms our natural antioxidant systems, brain fog, memory loss, DNA damage, and the onset or worsening symptoms of Alzheimer's, Parkinson's, multiple sclerosis (MS), Lewy body dementia, and autism ensue.

Intact (let's call them *fresh*) polyunsaturated fats are vulnerable to oxidation, but when they appear in their natural state, contained in whole foods, they are bundled with fat-guarding antioxidants like vitamin E. This is not the case when polyunsaturated fats appear in oils that have undergone heat and chemical processing. When these oils are extracted and used to create packaged foods, they represent one of the major toxins in our food supply.[5]

Sometimes these oils are where you'd expect to find them, like in commercial salad dressings and margarines. Other times, they are sneakier. Grain-based desserts like cookies and cakes, granola bars, potato chips, pizza, pasta dishes, bread, and even ice cream are among the top sources of oxidized oils in the diet.[6] They coat and comprise the "varnish" on breakfast cereals. "Roasted" nuts are covered with them (unless they explicitly say they are dry-roasted). And these oils are regularly served to us in restaurants, where processing, poor storage methods (being left out in a warm kitchen environment for months, for example), and heating and reheating make these highly susceptible fats go bad. Most restaurants now fry and sauté foods in them, reusing the oil over and over, further damaging them, and damaging you in the process. French fries? Shrimp tempura? Those delicious beer-battered chicken fingers? All are vehicles for these biochemically mutated oils, and for massive amounts of dangerous compounds called *aldehydes*.

Aldehydes are by-products of fat oxidation and have been found in elevated amounts in Alzheimer's-riddled brains. They may influence the susceptibility of proteins in the brain to cross-link and clump together, thereby forming the plaques that gunk up the brain and are characteristic of the disease.[7] These chemicals also serve as powerful toxins to the energy-generating mitochondria of the brain and spinal cord.[8] Aldehyde exposure (resulting from consuming rancid oils) directly impairs cells' ability to generate energy. This is pretty bad news for your brain, the chief energy consumer in your body.

Even after one polyunsaturated oil–rich meal, circulating markers of fat oxidation skyrocket by about 50 percent in young people, while a fifteen-fold increase in markers of rancid oils has been observed in older subjects.[9] Another study noted that arteries become instantly stiffer and less responsive to the demands of exercise after a similar meal. These fats, far removed from their natural form, fuel the underlying mechanisms of chronic disease, damaging your DNA, causing inflammation in your blood vessels, and raising your risk for several types of cancer.

These are the ominous oils to watch out for:

Canola oil	Safflower oil
Corn oil	Sunflower oil
Soybean oil	Rapeseed oil
Vegetable oil	Grapeseed oil
Peanut oil	Rice bran oil

The food industry's search for cheap oil that it could market to the American people resulted in a veritable rogue's gallery of deplorables. Sure, we eventually found out that trans fats were worse for our health than real butter could ever be, but our veil of ignorance continues to be exploited on butter-yellow tubs with labeling like "no hydrogenated oils," "non-GMO," and, of course, "organic." In reality, these wellness buzzwords only serve to obscure the few pennies' worth of mutated, rancid, heat-damaged Frankenfats that have been squeezed into a tub and sold for $4.99 in the premium health food section of the supermarket.

Cottonseed, canola, safflower, sunflower, and soy oils—all are bad news and are hidden virtually anywhere manufacturers can squeeze them. In all, our use of these oils has skyrocketed two hundred– to one thousand–fold in the last century (the latter figure being the case for soy), despite an overall 11 percent decrease in total fat consumption by adults in the United States between the years 1965 and 2011.[10] These oils now make up 8 to 10 percent of total caloric intake for Americans—up from almost zero at the turn of the century. While a handful of sunflower seeds or peanuts or a corn on the cob may be perfectly healthy, there is no safe level of consumption for any of these oils when industrially extracted from their original food sources and heated to high temperatures.

FAQ: I thought canola oil was healthy because it contains omega-3s?

A: Canola oil is highly processed. While it does contain a relatively high amount of omega-3s compared to other oils, omega-3s are even **more** vulnerable to oxidation than omega-6s. The processing of canola oil creates just as many oxidative by-products, including trans fats, which damage your blood vessels and your brain cells.[11] More on this later.

A Brain on Fire

We tend to think of our brains as unaffected by the goings-on in the rest of our bodies, but the problems associated with inflammation don't stay below the neck. Perhaps we don't think much about inflammation in the brain because it's invisible—it's not something we can feel with unmistakable certainty, as we can pain in an arthritic knee or an upset stomach, for example. But here's the cold, hard truth: our brains sit downwind of the activated immune system. Alzheimer's disease, Parkinson's, vascular dementia, MS, and chronic fatigue syndrome—these diseases can all in some way be likened to forest fires in the brain, frequently set off by a spark elsewhere in the body. But even before disease sets in, inflammation can rob us of our cognitive potential. If clear thinking is like coasting down a multilane highway with no traffic and all lanes open, inflammation creates lane closures and bottlenecks in the traffic.

Having evolved over millennia, a competent and highly adaptable immune system is vital to our survival—without one, even the most minute infection could lead to death. The immune system fights off these infections and is also the mechanism that engorges injured parts of the body with blood to help them heal—a sprained ankle, for example. The ensuing heat and redness (the *flame* in *inflamma-*

tion) is perfectly healthy—desired, even—under the conditions I just described. Unfortunately, our immune systems today are in a constant state of activation, not due to infectious threat but rather to what we're eating.

Whereas omega-3 fats like DHA and EPA are anti-inflammatory, omega-6 fats are the raw materials used in our bodies' inflammation pathways—the very same pathways that become activated when the body is under the assault of infection. While it is speculated that our ancestral diets incorporated these essential fatty acids in a ratio of roughly one to one, today we're consuming omega-6 fats and omega-3 fats in a ratio of twenty-five to one.[12] This means that every gram of omega-3 fat we consume is being washed down with a whopping 25 grams (or more) of omega-6. This kicks the aging process into high gear, accelerating the degenerative processes that underlie many of the chronic diseases burdening society today, and making you feel like crap all the while.

How might you use fat to your advantage? Aside from cutting out polyunsaturated oils from your diet (like grapeseed oil, frequently hidden in salad dressings, which has an omega-6–to–omega-3 ratio of seven hundred to one!), increase your consumption of foods that are naturally high in omega-3s. This can be achieved by sticking to wild fish, pastured eggs, and grass-fed or pasture-raised meats, which have more omega-3s and fewer omega-6s. If you don't like fish, or are unable to consume it two to three times per week, consider supplementing with high-quality fish oil (I'll provide tips for choosing one in chapter 12, but here's a hint: fish oil is the one place you *don't* want to skimp). One study from Ohio State University found that by simply taking a daily fish oil supplement that had 2085 milligrams of anti-inflammatory EPA per day, students were able to achieve a 14 percent reduction in one particular marker of inflammation. (This coincided with a 20 percent reduction in their anxiety.)[13]

ARIGATO . . . FOR THE ALZHEIMER'S?

The Japanese dietary pattern is known to include lots of vegetables and copious amounts of fish, the latter being a rich source of both DHA and EPA omega-3s. The country also enjoys low rates of Alzheimer's disease. However, when Japanese nationals move to the United States and adopt the inflammatory Western diet rich in polyunsaturated oils, factory-farmed meats, and refined carbohydrates, that protection seems to disappear: Alzheimer's rates among Japanese people living in the United States are more similar to those of Americans than to their relatives back home.[14]

Sane in the Membrane

Whether you're going over a presentation, doing your taxes, or deciding what to watch on Netflix, your thoughts are the end result of countless chemical (and electrical) reactions that occur across the quadrillion connections that neurons make with one another in your brain. And the success of these processes may come down to one vital, unsung hero of our cognitive function: the cell membrane.

Aside from forming protective barriers, membranes also provide the "ears" of neurons by cradling receptors for various neurotransmitters within. Neurotransmitters are chemical messengers, and there are dozens of them in the brain (you may have heard of the A-listers serotonin and dopamine, neurotransmitters associated with positive mood and reward). Much of the time, receptors for these messengers will sit under the membrane's surface, lying in wait for the right signal before bobbing up to the surface like buoys on the water.

A properly functioning neuron must have the ability to increase or decrease its sensitivity to outside signals, and it does this by increasing and decreasing the number of buoys allowed to the surface.

For this to occur, the cell membrane must possess the property of *fluidity*. This is true of most cells in the body, but it is especially important for neurons. If the nerve cell membrane is too rigid, receptor availability is impaired and can result in dysfunctional signaling, thus influencing our moods, behaviors, and memories.

The good news is that, as with inflammation, your diet directly affects neuronal membrane fluidity. Membranes are formed by substances called *phospholipids*, which are essentially the chemical structures that hold important building blocks like DHA in place in the cell membrane. When these structures are rich in DHA (from fatty fish, for example), membranes behave more fluidly, allowing the various receptors the ability to pop up to the surface of the cell membrane to "hear" the various messages from neurotransmitters. Unfortunately, omega-6 fats and omega-3 fats are like highly competitive soccer rivals, both vying for the same trophy—in this case the limited real estate in cell membranes.

In a diet where omega-3s and omega-6s are consumed in comparable amounts, the brain's ideal structural balance would be met. But today, because most of us overconsume omega-6 fats by an order of magnitude, we elbow out the omega-3 fats and enrich these phospholipid structures with omega-6 fats instead. This promotes a more rigid membrane, making it difficult for these important signaling receptors to surface.[15] When that happens, our mental health—and aspects of our intelligence—may suffer.

BDNF: THE ULTIMATE BRAIN BUILDER

Omega-3 fats and particularly DHA directly support the brain by increasing its supply of a protein called *brain-derived neurotrophic factor*, or BDNF for short. Dubbed "Miracle-Gro for the brain," BDNF is known not only for its ability to promote the creation of new neurons in the memory center of the brain but also for being a *bodyguard* to your existing brain cells, helping to ensure their survival. The astonishing power of BDNF can be seen when the protein is sprinkled on neurons in a petri dish—it causes them to sprout dendrites, the spiny structures required for learning, like a Chia Pet!

Having higher levels of BDNF bolsters memory, mood, and executive function in the short term, and is a powerful promoter of brain plasticity in the long term.[16] *Plasticity* is the term that neuroscientists use to describe the ability of the brain to change. In conditions where this characteristic is diminished, including Alzheimer's and Parkinson's, BDNF is also lower. The Alzheimer's brain may actually have half of the BDNF of a healthy brain, and raising it might slow progression.[17] Even depression may be a result of having lowered BDNF, and increasing it might improve symptoms.[18]

While exercise is one of the best overall ways to boost this powerfully protective growth hormone, consuming omega-3 fats, and particularly DHA, is among the best dietary means that we know of. DHA is so important for building a healthy brain that researchers believe it was access to this special fat that allowed our early hominid brains to reach their current size. This may explain why the consumption of fish, leading to higher blood levels of omega-3 fats including DHA, is correlated with greater total brain volume over time.[19] But don't write off DHA's usual sidekick, EPA: inflammation is a well-known BDNF brain drain, and EPA is a powerful inflammation quencher.

Un-jamming Brain Gridlock . . . with Fat

Throughout my childhood, I had difficulties that seem like common complaints today: I was easily distracted and had a hard time sitting still and focusing on my classwork. As a result, I struggled to get good grades. At one point, my school's guidance counselor even suggested to my parents that they send me to see a psychologist. (Look at me now, Mrs. Capello!)

Grievances aside, the troubles I had fall under the domain of executive function, a broad set of cognitive abilities that includes planning, decision making, attention, and self-control. Executive function is so far-reaching in day-to-day life that some experts think it's more important to success than IQ or even inherent academic talent.[20] And luckily, research has highlighted a role for dietary fat in optimizing it.

Like all areas of cognitive function, executive function relies on the healthy functioning of neurotransmitters. As such, it may be particularly affected by omega-6–to–omega-3 imbalances. Researchers have observed that children with lower intakes of omega-6 fats perform significantly better in regard to their executive abilities.[21] And for children with attention-deficit/hyperactivity disorder (ADHD), which is often described as an executive function problem,* as well as for typically developing children, attention has been shown in some studies to improve with omega-3 supplementation.[22] (Were the margarine and grain oils I grew up eating directly responsible for my issues? I'll never know for sure—but it wouldn't be hard to believe.)

When it comes to shifting our fat intake toward a healthier state, anytime is the right time—even if that means simply adding a fish oil supplement, according to a trial out of Berlin's Charité Hospital.[23] In this study, adults were given daily omega-3 supplements containing 1320 milligrams EPA and 880 milligrams DHA. After twenty-six weeks, the researchers found that subjects taking the omega-3 supplements displayed executive function enhanced by 26 percent over the placebo group, who actually saw a slight decline

---

* The modern "problem" of ADHD may be more a consequence of brains wired for novelty and exploration clashing with routine jobs and one-size-fits-all education, a theory that I will revisit in chapter 8.

in their cognition. They also showed an increase in gray matter volume and "superior white matter structural integrity." Think of white matter as the interstate highway system of the brain, allowing data to be shuttled between different regions at express-lane speeds. In this study, omega-3 supplementation seemed to act like an infrastructural reinforcement team, smoothing out the potholes on the highway and even adding extra lanes.

Helping you perform better is one thing, but might adding more omega-3s to your diet help if you were one of the 450 million people across the globe who suffer from some kind of mental illness? That's the question University of Melbourne researchers asked when they gave a daily dose of fish oil to people in their teens and early twenties with a history of psychotic symptoms. (Using fish oil as a preventative or therapeutic approach is also very appealing because it doesn't carry a stigma the way antipsychotics do.)

Each subject in the trial was given 700 milligrams of EPA and 480 milligrams of DHA per day. Over three months, researchers found that, compared to the placebo, the fish oil group displayed significantly fewer psychotic episodes.[24] Even more impressive: the improvement in symptoms seemed to persist when doctors assessed subjects' mental health *seven years later*—only 10 percent converted to full-on psychotic disorders compared to 40 percent in the placebo group (a fourfold risk reduction). Patients were also significantly higher functioning, and needed less medication to manage their symptoms.[†]

Is fish oil a cure-all for mental health? No, sadly. But this research does provide further evidence that our diets have become disharmonious with the needs of our brains—and by correcting the imbalance, we may reap significant benefits.

[†] Previously, omega-3s had shown mixed results in adults with psychotic disorders, but this study offers evidence that starting treatment earlier may be more effective.

FURANS—YOUR BRAIN'S SLEEPER AGENT?

The late Austrian chemist Gerhard Spiteller, the first scientist to ring the alarm bell on the dangers of processed polyunsaturated oils, was studying fish oils when he made a fascinating observation. He noticed that concentrated sources of omega-3s were always accompanied by a type of fat called a *furan fatty acid*, or F-acid. Made by algae and plants, these F-acids are incorporated into fish oil when fish eat algae. (Another known F-acid source is organic, grass-fed butter.)[25] Once consumed by us, they travel alongside omega-3s and 6s and other fats in a cell membrane where they scavenge and neutralize nearby free radicals generated by polyunsaturated fats or other oxidative stress.

Japanese researchers saw the power of these mysterious fats when they studied the potent anti-inflammatory effects of the New Zealand green-lipped mussel. Curious about the much lower rates of arthritis in the coastal, mussel-chomping Maori population when compared to their inland-dwelling counterparts, the scientists compared the F-acid-containing mussel extract to EPA-rich fish oil and found it was almost *one hundred times* more potent than the EPA at reducing inflammation!

How do F-acids achieve this? They contain what's called a *resonance structure*. A resonance structure may sound like the crystal that powers a lightsaber or Iron Man's suit, but it's actually even cooler: these chemical firefighters knock out free radicals, and then stabilize themselves to end the destructive chain reaction. They're so good at it, F-acids *may* be your brain's silent guardian molecules, sniping free radicals like a boss while allowing omega-3s to take all the credit.

Let's take a moment of pause, however, before we try to make F-acids the next big supplement craze. The discovery of these benevolent free radical fighters is an argument *against* trying to break the value of whole foods down into individual micronutrients. We've coevolved with our food, and trying to optimize our infinitely complex bodies by cherry-picking nutrients may be the ultimate exercise in hubris. F-acids are the perfect case in point: pharmaceutical companies have been trying to distill and extract purer and purer concentrations of EPA omega-3s from fish to create super-potent fish oil, but they don't always show the expected anti-inflammatory benefit. Could this be because these super-delicate yet powerful furan fats are destroyed in manufacturing? This is why we're always in favor of whole foods over supplements—even the supplements we recommend!

ALA—The Plant Omega-3

I've briefly mentioned another common omega-3: plant-based alpha-linolenic acid, or ALA, found in seeds and nuts like flaxseeds, chia seeds, and walnuts. In our bodies, ALA needs to be converted to DHA and EPA to be used, but this is a very inefficient process, and what limited ability we possess further declines with age.[26]

Healthy young men convert an estimated 8 percent of dietary ALA to EPA, and 0 to 4 percent to DHA. In fact, the conversion of ALA to DHA is so limited in men that consuming more ALA (from flaxseed oil, for example) may not increase DHA in the brain at all. Women, on the other hand, are approximately 2.5 times more efficient at converting ALA, an ability thought to be facilitated by estrogen to support the needs of future childbearing. Unfortunately, the capacity to create DHA from ALA may partly decline as a result of menopause, perhaps playing a role in the increased risk that women face for both Alzheimer's disease and depression.[27]

Factors other than gender influence the conversion of plant-based ALA to DHA and EPA. People of European origin who possess "newer" genes (they just don't make 'em like they used to) may have reduced conversion abilities compared to those of African descent—it's possible that the ability to convert plant forms of ALA became relegated with the increasing availability of more reliable sources of omega-3s from meat, fish, and eggs.[28]

Ironically, and adding to the considerable consequences of polyunsaturated oil consumption, the enzymes that convert ALA to EPA and DHA also convert linoleic acid, the predominant omega-6 fat in the diet, to its usable pro-inflammatory form (called *arachidonic acid*). These benevolent worker chemicals are indifferent to our needs—they just convert what we feed them, and today, we're feeding them mostly omega-6s. In the case of people who get little preformed EPA and DHA and lots of omega-6s from their diets (vegans who consume lots of processed foods, for example), the brain may actually become omega-3 deficient for this reason.

To eliminate the guesswork when it comes to nourishing your brain with EPA and DHA, I suggest the "set it and forget it" method: be vigilant in your avoidance of polyunsaturated oils—corn, soy, canola, and other grain and seed oils—and ensure that you're getting *preformed* EPA and DHA from whole-food sources like fish (wild salmon and sardines are great, low-mercury choices), pastured or omega-3 eggs, and grass-fed beef. On days that you are unable to get your dose of preformed EPA and DHA, supplemental fish, krill, or plant-based algae oil may help. Once you cover those bases, ALA from whole-food sources like walnuts, flaxseeds, or chia seeds is a great addition.

Monounsaturated Fats: Your Brain's Best Friend

As with polyunsaturated fats, the brain is rich in monounsaturated fats, which form the brain's myelin sheath. This is the protective coating that insulates neurons and allows for speedy neurotransmission. However, unlike polyunsaturated fats, monounsaturated fats are chemically stable. Oils composed primarily of these fats not only are safe to consume but seem to have a number of positive effects in the body. Some common sources of monounsaturated fat include avocados, avocado oil, and macadamia nuts, and the fat content of wild salmon and beef is nearly 50 percent monounsaturated. But perhaps the most famous source of monounsaturated fat is extra-virgin olive oil.

In Mediterranean countries such as Greece, southern Italy, and Spain—where rates of neurodegenerative diseases like Parkinson's and Alzheimer's are lower—extra-virgin olive oil is the ultimate sauce, used liberally on steak, beans, vegetables, bread, pizza, pasta, and seafood, in soups, and even in desserts. My friend Nicholas Coleman, chief oleologist at New York City's Eataly, painted the picture for me: "They don't drizzle olive oil; they pour it on." Mediterraneans even cook with it—contrary to popular belief, EVOO retains much of its nutritional value even under extreme conditions.[29] (That being said, it's still better to save high-heat cooking for saturated fats, which are the most chemically stable—and which we'll cover next.)

The so-called Mediterranean diet is often cited by epidemiologists (scientists who study health and disease in large populations and make associations based on the data that they collect) as being the most protective large-scale dietary pattern against cardiovascular disease and neurodegeneration, and it's been shown that higher adherence to the Mediterranean style of eating leads to not only better long-term health outcomes (including a robust risk reduction for developing dementia), but bigger brains as well.[30] But as I've mentioned, the major limitation of epidemiological studies is that they are based on observation, making it impossible to pinpoint which aspects of the diet are causally involved in such benefits. To bridge this gap and look specifically at the effect of foods rich in monounsaturated fat on cognitive performance, scientists in Barcelona began a trial that pitted a low-fat diet (still widely recommended) against two versions of the higher-fat Mediterranean diet.[31]

One of the two experimental Mediterranean diets was supplemented with tree nuts like almonds, hazelnuts, and walnuts—all great sources of monounsaturated fat. The other experimental diet was supplemented with *even more* extra-virgin olive oil. In the high olive oil group, participants were given a liter to consume per week. Just to put that in perspective, one liter of olive oil contains more than 8,000 calories—more than half a week's worth of calories for an adult male! Both groups—those that adhered to the diet supplemented with added nuts and those supplementing with added olive oil—not only retained but *improved* their cognitive function after six years, with the olive oil group coming out slightly ahead. The low-fat control group exhibited steady decline.

Get to know the grassy, peppery taste of a good EVOO (preferably organic) by swooshing it to the back of your throat—and taste it often! Stock your kitchen with extra-virgin olive oil, and use it in low- to medium-heat cooking, as a sauce on eggs, vegetables, and fish, and in all of your salads.

Saturated Fats: Stable and Able

Saturated fats are essential to life—they provide support to your cell membranes and serve as precursors to a variety of hormones and hormonelike substances. Saturated fat is the most abundant type of fat in human breast milk—arguably nature's ideal food for a newborn.[32]

Usually solid at room temperature, saturated fat is most commonly found in full-fat dairy like cheese, butter, and ghee, meats like beef, pork, and chicken, and even certain fruits like coconut and olives. (Extra-virgin olive oil is nearly 15 percent saturated fat.)

Saturated fats have had a lot of bad press in recent years, having become vilified as the "artery-clogging" fat. Quite literally, these are the fats our mothers warned us about. But unlike the toxic fats we've traded them in for (grain and seed oils like canola, corn, and soybean oil), saturated fats are the most chemically stable and the most appropriate to use for higher-heat cooking. Welcoming saturated fats (such as coconut oil, grass-fed butter, and ghee) back to the kitchen is a biologically relevant, real-world application that may have a major benefit on your health.

A Fat Framed?

As a nutrient, saturated fat is not inherently unhealthy or healthy. Its role in your health is dependent on a few questions, such as: Do you eat a lot of sugar? Is your diet heavy in processed foods? Do you consider ketchup a vegetable? This is because saturated fat can magnify the deleterious effects of a high-carbohydrate, low-nutrient diet. (There's also a question of genes, which I will explore in chapter 5.)

Unfortunately, ultra-processed convenience foods tend to be high in sugar and refined carbohydrates, and they are often combined with equal amounts of saturated fat. Picture hamburgers on white-flour buns, cheesy pizzas, creamy pasta dishes, deluxe nachos, burritos, ice cream, and even the seemingly innocuous bagel with butter. These foods now make up 60 percent of the calories consumed in the United States and are highly damaging to our health.[33]

Some research suggests that the combination of carbs and fat in a given meal can induce a temporary state of insulin resistance, a form of metabolic dysfunction that increases inflammation and fat storage. (I'll describe exactly how this affects the brain in upcoming chapters.) That our bodies become confused when high amounts of saturated fat and carbohydrates are consumed together should not come as a surprise. After all, you would have a hard time finding foods in nature that contain both saturated fat *and* carbohydrates. Fruit is mostly pure carbohydrate and fiber, and low-sugar fruits like avocado and coconut contain ample fat but very little carbohydrate. Animal products are usually pure fat and protein. And vegetables, whether starchy or fibrous, are usually free of fat. Dairy would be the one exception, where saturated fat and sugars are combined—which may help it serve its evolutionary purpose of helping a young animal put on weight. Otherwise, only modern foods regularly marry saturated fat and carbohydrates, combined usually with the intent of promoting overconsumption.

SATURATED FATS IN YOUR BLOOD

Blood levels of saturated fat have been linked with increased risk for dementia, but how do those fats get there in the first place?[34] "It is commonly believed that circulating fatty acids reflect dietary intake, but the associations are weak, especially for SFA (saturated fatty acids)," wrote researchers from Ohio State University in *PLOS ONE*, seeking to answer this very question.[35] They found that two of the saturated fats linked with dementia, stearic and palmitic acids, did not become elevated in the blood even when their subjects consumed them in amounts as high as 84 grams per day—the equivalent of nearly eleven tablespoons of butter! On the other hand, the highest circulating saturated fats were measured after the subjects consumed a high-carbohydrate diet, while eating fewer carbohydrates led to lower circulating levels. As it turns out, most of the body's circulating levels of saturated fats originate in the liver, where they are produced in response to carbohydrates—a process called *lipogenesis*, or fat creation. Other studies have demonstrated similar results, proving that our bodies are dynamic chemistry labs that don't always follow simple logic—a fact often used to sell food products, pharmaceutical drugs, or general misinformation.[36]

Saturated Fat and the Brain: Friends or Foes?

When it comes to the impact that saturated fat has on the brain, finding truthful answers can be tricky. Close inspection of many animal studies nearly universally reveals that what is reported as a "high-fat diet" for the animals is, in reality, a toxic slurry of sugar, lard, and soybean oil.[‡] This might trace back to a basic labeling oversight—lab suppliers of rat chow often label diets meant to mimic the Standard American Diet as simply "high fat."

Don't get me wrong: animal studies like these are incredibly valuable. Thanks to these studies, we have some clues as to why people who adhere more closely to the high-sugar, high-fat Standard American Diet tend to have smaller hippocampi—the structure in the brain that processes our memories.[37] These studies also tell us that the combination of sugar and saturated fat (common in fast food) can drive inflammation and drain BDNF from the brain.[38]

The problem is, this nuance is often lost when the media reports on these findings, resulting in misleading headlines such as "How a High-Fat Diet Could Damage Your Brain"—which was the title of a widely circulated article posted on one well-known publisher's site.[39] (The food used in the mouse study that the article was reporting on was 55 percent saturated fat, 5 percent soybean oil, and 20 percent sugar.) Unless readers went out of their way to find the original study, assuming they'd even have access to it and didn't glaze over from the jargon, they could easily interpret this as a strike against high "healthy fat" diets—those that are low in processed carbohydrates and polyunsaturated oils and high in omega-3 fats, nutrient-rich vegetables, and the relatively small amount of saturated fat found in properly raised animals' products.

The question that remains is how much saturated fat should be consumed in a brain-optimal diet. While the evidence warning us to avoid saturated fat is, and has always been, shaky at best, there is also scant evidence to suggest that *chasing* saturated fat has any benefit to the brain (unlike, for example, monounsaturated fat, which is the primary fat in extra-virgin olive oil). While the details are

---

‡ Sometimes experimental high-fat diets might even include *trans* fats. This oversight makes little sense considering that man-made trans fats are highly toxic, possessing clear deleterious effects on cognitive health.

still being unraveled, you can rest assured that what's good for your body is also very likely good for your brain. What we're beginning to learn is that nutritionally poor Westernized diets that are rich in processed polyunsaturated oils and rapidly digestible carbohydrates are the true culprits in not only cardiovascular disease but obesity and type 2 diabetes—and, as research is now making clear, brain disease as well.

For these reasons, I place no restriction on the consumption of saturated fats *when they are contained in whole foods*, or when they are used for occasional higher-heat cooking. (The main *oil* in your diet should always be Genius Food #1—extra-virgin olive oil.)

Trans Fats: A Fat to Be Feared

Trans fats are unsaturated fats that behave in some ways like saturated fats. One naturally produced trans fat, conjugated linolenic acid (CLA), is found in the milk and meat products of grass-fed animals, and is believed to be very healthy, associated with better metabolic and vascular health and reduced cancer risk. But natural trans fats are relatively rare in the modern human diet.

The bulk of trans fats consumed by humans are the result of industrial manufacturing. These man-made trans fats are not just bad; they're *Darth Vader–meets–Lord Voldemort* bad. They begin life as polyunsaturated oils (which can freely pass the blood-brain barrier), and are pumped with hydrogen. You can see these on food packages if you look for hydrogenated or partially hydrogenated oils. This process makes them behave more like saturated fats, becoming solid at room temperature. Food manufacturers like this for two reasons: it allows them to add a rich, buttery texture to foods using cheap oils, and it extends the shelf life of those foods. As such, these fats are commonly found in packaged foods, cakes, margarine, nut butters (where they prevent oil separation), and even some vegan "cheese" spreads with otherwise healthy-looking packaging.

Man-made trans fats are highly inflammatory, promoting insulin resistance and heart disease (they can raise total cholesterol while lowering protective HDL). A recent meta-analysis (a study of studies) found that consumption of trans fats was associated with a 34 percent increased risk of all-cause mortality, meaning early death by any cause.

In terms of the brain, trans fats may be particularly damaging. Remember earlier when I told you about the value of membrane fluidity? Trans fats can integrate themselves into your neuronal membranes and stiffen them like a corpse with rigor mortis. This makes it much more difficult for neurotransmitters to do their jobs, and for cells to receive nutrients and fuel. Studies have also linked trans-fat consumption to brain shrinkage and sharply increased risk for Alzheimer's disease—two things that you certainly don't want.[40] But, even in healthy people, consuming trans fats has been associated with significantly worse memory performance. A study published in 2015 found that for every additional gram of trans fat participants ate, their recall of words that they had been asked to remember dropped by 0.76 words.[41] Those who ate the most trans fats recalled twelve fewer words than those who consumed no trans fats at all.

Think you're safe just by avoiding hydrogenated oils? The mere processing of polyunsaturated fats *creates* trans fats—researchers have found small amounts of them lurking within the bottles of many commonly sold cooking oils. Even organic, expeller-pressed canola oil is as much as 5 percent trans fat. We consume on average about 20 grams of canola or other vegetable oil per person per day. That's 1 gram of trans fat right there.

By avoiding corn, soy, and canola oils (and products that are made with them) as I've outlined earlier in the chapter, and any oil that has been "hydrogenated" or "partially hydrogenated," you are covering your bases to ensure that not a trace of man-made trans fat enters your system.

Fat: The Nutrient Ferry

A final but incredibly important benefit of adding more good fats to your diet (in the form of fat-rich foods like eggs, avocado, fatty fish, and extra-virgin olive oil) is that fats facilitate the absorption of critical fat-soluble nutrients like vitamins A, E, D, and K, as well as important carotenoids like beta-carotene. These nutrients have wide-ranging effects on the body, from protecting against DNA damage to guarding the fats that are already present in your body and brain against aging.

Carotenoids—the yellow, orange, and red pigments abundant in carrots, sweet potatoes, rhubarb, and particularly in dark leafy greens like kale and spinach—have been identified as powerful brain boosters (you can't see them in dark leafy greens because they're disguised by the green pigment of chlorophyll—but they're there). Among these, lutein and zeaxanthin in particular have been linked with greater neural efficiency and "crystallized intelligence," or the ability to use the skills and knowledge one has acquired over a lifetime.[42]

TURBOCHARGE YOUR BRAIN WITH CAROTENOIDS

It's been known for some time that carotenoids play an important role in protecting the eyes and brain from aging, but they may also *speed up* your brain. In a clinical trial, University of Georgia researchers gave sixty-nine young, healthy male and female students supplements containing lutein and zeaxanthin—two carotenoids abundant in kale, spinach, and avocado—or a placebo for four months. The subjects who received the lutein and zeaxanthin saw a 20 percent boost in visual processing speed, measured by the retina's automatic reaction to a stimulus. Processing speed is important because it's the pace at which you take information in, make sense of what you are perceiving, and begin to respond. Faster visual processing tends to correlate with better sports performance, reading speed, and executive function, and reduced processing speed is a central and early feature of cognitive decline. Researchers wrote of this impressive increase, "[This] is significant since young healthy subjects are typically considered to be at peak efficiency and might be expected to be most resistant to change." They continued: "It can be generally remarked that improving diet is not simply to prevent acquired or deficiency disease, but rather to optimize function throughout life." I couldn't agree more!

These nutrients *require* fat to be piggybacked into your body's circulation, to the degree that when a salad is consumed, the absorption of carotenoids is negligible unless eaten with a source of fat.[43] A generous splash of extra-virgin olive oil is an excellent choice, or simply add a few whole eggs to your salad. In a Purdue University study, participants who added three whole eggs to their salads increased their absorption of carotenoids by three- to eightfold compared to when no eggs were added.[44] If eggs aren't your thing, add some avocado, and know that in doing so you are reaping the astonishing benefits of fat-soluble, brainpower-boosting nutrients like carotenoids.

So there you have it—a solid understanding of the role of fat in your body. Many generations of families were led astray when it came to our most important nutrient, but we now know the importance of the right fats for the brain. Once I learned this, my diet changed dramatically, and satiating, nourishing, and *safe* foods that were once off-limits have become irreplaceable staples in my diet.

But we aren't in the clear just yet. Cognitive catastrophe (and salvation) merely *begins* with fat. Up next, the ultimate harbinger of brain destruction.

FIELD NOTES

- Polyunsaturated fats, vulnerable to oxidation, can either be your best friend or worst enemy. Avoid grain and seed oils like corn and soy, as well as fried foods that use recycled vegetable oils.
- Make your own salad dressing. You really don't want 200 calories of suspect polyunsaturated fats served with your healthiest meal of the day. Store-bought and even restaurant-made salad dressings can be the worst offenders. Restaurants routinely swap out or dilute EVOO with canola oil or worse—mystery "vegetable oil"!
- Restaurant food is almost always a gamble, so look the proprietors in the eye and ask them what oils they cook with.
- If you are unable to get three-plus servings of fatty fish per week (wild salmon and sardines are mega-concentrated sources of omega-3), consider taking a fish oil supplement; or, if you're a vegan, opt for algae oil.
- Extra-virgin olive oil should be your main dietary oil.
- Saturated fat from whole-food sources is healthful in the context of a diet that is devoid of sugar, low in carbohydrates, and high in fiber, omega-3s, and essential nutrients from plant foods.
- Trans fats are a dietary devil. Avoid anything with hydrogenated oil or processed polyunsaturated oils, which are at least 5 percent trans fat even without hydrogenation.
- Certain nutrients in vegetables are not absorbed unless in the presence of fat—salads and vegetables therefore should always include a healthy fat source.

GENIUS FOOD #2

AVOCADOS

Avocados are an all-in-one Genius Food—the perfect food to protect and enhance your brain. To start, they have the highest total fat-protecting capacity of any fruit or vegetable. This is good news for your brain, which is not only the fattiest organ in your body, but also a magnet for oxidative stress (a major driver of aging)—a consequence of the fact that 25 percent of the oxygen you breathe goes to create energy in your brain! Avocados are also rich in different types of vitamin E (a characteristic not many supplements can claim), and they are a potent repository for the carotenoids lutein and zeaxanthin. You may recall from chapter 2 that these pigments can boost your brain's processing speed, but that they rely on fat to be properly absorbed. Conveniently, avocados are an abundant source of healthy fats.

Today there is an epidemic of vascular disease, not only in the form of heart disease, but as vascular dementia, which is the second most common form of dementia after Alzheimer's. Potassium works with sodium to regulate blood pressure and is essential for vascular health, but today we tend to consume insufficient amounts of potassium. In fact, scientists believe that our hunter-gatherer ancestors consumed four times as much potassium as we do today, which may explain why hypertension, stroke, and vascular dementia are now so common. By providing twice the potassium content of a banana, a whole avocado is the perfect food to nurture the brain's estimated four hundred miles of microvasculature.

Finally, who needs fiber supplements (or cheap, industrially produced morning cereals) when you can eat an avocado? One whole medium avocado contains a whopping 12 grams of fiber—food for the hungry bacteria that live in your gut, which will ultimately pay their rent in the form of life- and brain-sustaining compounds that reduce inflammation, enhance insulin sensitivity, and boost growth factors in the brain.

How to use: I try to eat a half to a whole avocado every day. You can enjoy avocados simply sprinkled with a little sea salt and extra-virgin olive oil. They may also be sliced and added to salads, eggs, smoothies, or my Better Brain Bowl (see page 333).

Pro tip: Avocados are known for taking a long time to ripen, and only a day or two to foul. To keep extra avocados from going bad, pop them in the fridge once ripe, and take them out when ready to eat. You, 1; avocados, 0!

CHAPTER 3

OVERFED, YET STARVING

A human being should be able to change a diaper, plan an invasion, butcher a hog, conn a ship, design a building, write a sonnet, balance accounts, build a wall, set a bone, comfort the dying, take orders, give orders, cooperate, act alone, solve equations, analyze a new problem, pitch manure, program a computer, cook a tasty meal, fight efficiently, die gallantly. Specialization is for insects.

—ROBERT A. HEINLEIN

Let us think back to a time before food delivery apps and diet gurus, when "Trader Joe" was the guy guarding the only salt lick in a hundred-mile radius and "biohacking" was something you did to a fresh kill with a sharpened stone. Government diet recommendations (or governments, for that matter) wouldn't arrive on the scene for millennia, so you'd have to make do, as your ancestors did, with intuition and availability. As a forager, your diet would consist of a diverse array of land animals, fish, vegetables, and wild fruits. The chief calorie contributor would by and large be fat, followed by protein.[1] You might consume a limited amount of starch, in the form of fiber-rich tubers, nuts, and seeds, but concentrated sources of digestible carbohydrate were highly limited, if you had access to them at all.

Wild fruits, the only sweet food available to ancestral you, looked and tasted much different from the domesticated fruits that would line supermarket shelves eons later. You likely wouldn't even recognize them when placed next to their contemporary counterparts, a contrast almost as stark as a Maltese lapdog standing next to its original ancestor, the gray wolf. These early fruits would be small, taste a fraction as sweet, and be available only seasonally.

Then, approximately ten thousand years ago, a hairpin turn in human evolution occurred. In the blink of an eye, you went from a roaming tribal forager subject to the whims of season to a settler with planted crops and farmed animals. The invention of agriculture brought to your family—and the rest of humanity—what was a previously inconceivable notion: the ability to produce a surplus of food beyond the immediate needs of daily subsistence. This was one of the major "singularities" of human existence—a paradigm shift marking a point-of-no-return entry into a new reality. And in that new reality, though we procured quantities of foods that would feed many people cheaply and fuel global population growth, individual health took a downward turn.

For hundreds of thousands of years prior, the human diet was rich in an array of nutrients spanning diverse climes, but this micronutrient and geographic diversity disappeared when every meal became based on the handful of plant and animal species that we were able to cultivate. Starvation was less of an immediate threat, but we became slaves to single crops, making nutrient deficiencies more prevalent. The dramatic increase in the availability of starch and sugar (from wheat and corn, for example) created tooth decay and obesity, a loss of height, and a decrease in bone density. By domesticating animals and crops, we inadvertently domesticated ourselves.

The advent of agriculture fed a vicious spiral of behavioral demands that changed the very nature of our brains. A hunter-gatherer had to be self-sufficient, but the post-agriculture world favored specialization. Someone to plant the wheat, someone to pick it, someone to mill it, someone to cook it, someone to sell it. While this process of hyper-specialization eventually led to the industrial revolution and all its conveniences like iPhones, Costco, and the Internet, these modern trappings came with a flip side. Fitting an ancient brain into a modern environment may be like fitting a square peg into a round hole, as evidenced by the millions of Americans on antidepressants, stimulants, and drugs of abuse. A person with ADHD, whose brain thrives on novelty and exploration, may have been the ultimate hunter-gatherer—but today this person struggles with a job that requires repetition and routine (the authors can—*ahem*—relate).

The confluence of this dietary shift and the relegation of our cognitive duties caused our brains to lose the volumetric equivalent of a tennis ball in a mere ten thousand years. Our ancestors from five hundred generations ago would have lamented our restrictive existences, and then apologized to us for engineering our cognitive demise. Forget about leaving the next generation with lower standards of living, student debt, or environmental destruction—our ancestors were so successful that they left us with *smaller brains.*

We didn't know it at the time, but in one fell swoop we had turned our backs on the diet and lifestyle that created the human brain, and adopted one that shrank it.

Energy Dense, Nutrient Poor

Given the obesity epidemic and the amount of food Americans and others around the globe routinely throw away (even slightly misshapen fresh vegetables get tossed out so your supermarket-going experience is as aesthetically pleasing as possible), it may surprise you to know that our bodies are still somehow . . . starving.

Have you ever wondered why so many packaged goods now have to be "fortified" with vitamins? There are more than fifty thousand edible plant species around the world—plants that provide a bevy of unique and beneficial nutrients that we consumed as foragers. And yet today, our diets are dominated by three crops: wheat, rice, and corn, which together account for 60 percent of the world's calorie intake. These grains provide a source of cheap energy, but are relatively low in nutrition. Adding in a few cents' worth of (usually synthetic) vitamins is the dietary equivalent of putting lipstick on a pig.

MICRONUTRIENTS GONE MIA

Potassium	Supports healthy blood pressure and nerve signals
B vitamins	Support gene expression and nerve insulation
Vitamin E	Protects fatty structures (like brain cells) against inflammation
Vitamin K_2	Keeps calcium out of soft tissues like skin and arteries and in bones and teeth
Magnesium	Creates energy and facilitates DNA repair
Vitamin D	Anti-inflammatory, supports a healthy immune system
Selenium	Creates thyroid hormones and prevents mercury toxicity

The above list covers only some of the essential nutrients lost to the modern diet. In total, there are roughly forty minerals, vitamins, and other chemicals that have been identified as essential to our physiology and are readily contained in the whole foods we're not eating.[2] As a result, 90 percent of Americans now fall short in obtaining adequate amounts of at least one vitamin or mineral.[3]

To complicate matters, nutrient intake guidelines are set only to avert population deficiencies. This means that even when we check all the institutionally recommended boxes, we may still be handicapping our bodies in serious ways. The recommended daily allowance (RDA) of vitamin D, for example, is meant only to prevent rickets. But vitamin D (generated when our skin is exposed to the sun's UVB rays) is a steroid hormone that affects the functioning of nearly one thousand genes in the body, many involved in inflammation, aging, and cognitive function. In fact, a recent University of Edinburgh analysis found low vitamin D to be a top driver of dementia incidence among environmental risk factors.[4] (Some researchers have argued that the RDA for vitamin D should be at least ten times higher than it currently is for optimal health.)[5]

When our bodies sense low nutrient availability, what's available will generally be used in processes that ensure our short-term survival, while long-term health takes a back seat. That's the theory initially proposed by noted aging researcher Bruce Ames. Dubbed the "triage theory" of aging, it's sort of like how a government may choose to ration food and fuel during wartime. In such cases, more immediate needs such as food and shelter might take priority, whereas public education would become a casualty. In the case of our bodies, loftier repair projects can become an afterthought to basic survival processes, all while pro-inflammatory processes run amok.

The downstream effects of magnesium deficiency may be the perfect example of such reprioritization. This is because magnesium is a mineral required by more than three hundred enzymatic reactions in the body with duties ranging from energy creation to DNA repair. If it is constantly shuttled into short-term needs, DNA repair takes a back seat. This effect is almost certainly magnified when we consider that nearly 50 percent of the population doesn't consume adequate amounts of magnesium, second in deficiency rates only to vitamin D, and yet it is easily found at the center of chlorophyll, the energy-generating molecule that gives dark leafy greens their color.[6]

Research has validated that inflammation wrought by nutrient scarcity is strongly linked with accelerated brain aging and impaired cognitive function.[7] Robert Sapolsky, author of *Why Zebras Don't Get Ulcers*, may have said it best when describing the similar reshuffling of priorities that occurs during stress: the body holds off on long-term projects until it knows that there will *be* a long term. After all, the major consequences of damaged DNA—a tumor, for example, or dementia—won't get in your way for years, decades even . . . but we need energy *today*.

Sugar and Carbs 101

One could argue that the principle shift from prehistory to modernity was to promote concentrated sources of carbohydrates from cameo appearances in our diets to starring roles. The most concentrated source of carbohydrates is refined sugar, which is now added to everything—from seemingly innocuous juices, crackers, and condiments to more blatant offenders like soft drinks. Even when we try our best to avoid these simple sources of carbohydrates, they can be hidden most inconspicuously. Anti-obesity researcher and crusader Robert Lustig has identified fifty-six unique terms food manufacturers use to disguise sugar—making it difficult, if not altogether impossible, to spot added sugar on ingredient lists unless you are the most diligent of sleuths. Here are just a few of the many names for sugar: cane juice, fructose, malt, dextrose, honey, maple syrup, molasses, sucrose, coconut sugar, brown rice syrup, fruit juice, lactose, date sugar, glucose solids, agave syrup, barley malt, maltodextrin, and corn syrup.

But it's not just overt forms of sugar that have achieved prominence in the modern diet. Grains like wheat, corn, and rice, tubers like potatoes, and modern sweet fruits are all cultivated for maximum yield of starch and sugar. Though these starches don't look or taste like sugar, they are simply chains of glucose, stored in energy-dense tissue in the seeds of plants. (At this point you may be wondering if this book is going to banish these forms of food from your life forever, and the answer is no. In later chapters, we'll demonstrate how to consume starchy foods with higher-energy density in a way that benefits you as opposed to making you fat and sick.)

Scientists believe that in the preagricultural past, we were consuming close to 150 grams of fiber a day. Today, we're eating more concentrated carbohydrates than ever before, and getting a meager 15 grams of fiber—on a good day. Critics of the ancestral diet often point to the probable consumption of ancient grains in the pre-agricultural diet, but regardless of the exact percentage, they were clearly accompanied by massive fiber content—a dramatic and critically important contrast to the calorie-dense, processed sources we have today.

It's important to be aware of the ease with which our bodies can break a starch into its constituent sugar molecules. This conversion process doesn't even wait for you to swallow—it begins in your mouth, thanks to an enzyme in saliva called *amylase.* (If you're like me, you learned this in your ninth-grade biology class. Allow a starch to linger in your mouth and you'll taste the sweetness as the starches begin to break down to their constituent sugars right on your tongue.) In fact, even before you take your first bite (or sip) of food, just *looking* at what you're going to eat stimulates the production of the storage hormone insulin, so that it can be ready to dispose of the oncoming sugar deluge.

Insulin's main job is to quickly shuttle sugar molecules out of your blood and into your fat and muscle tissue. By the time sugar makes a quick pit stop in your stomach and hops on the ten-minute subway ride to your bloodstream, your body's endocrine (hormone) system is already in full-on energy storage mode. But energy storage is just one part of the story—this process is also responsible for controlling the damage caused by having too much sugar in your blood.

The human body likes stability. It goes to great lengths to keep your body temperature within a narrow range (hovering around 98.6°F) at all times, and the same can be said for your blood sugar levels. Your entire circulating plasma volume (about five liters of blood) contains just *one single teaspoon* of sugar at any given time. This may cause you to look at your food in a different light, perhaps thinking twice before reaching for that glass of orange juice, which contains six times your body's circulating blood sugar in just a single cup. Or that delicious cranberry muffin beckoning you from the office kitchen, containing seventeen times the amount of sugar—dumped nearly instantly into your bloodstream upon consumption.

Okay, so what? Eat the sugar, insulin gets it out of the bloodstream—no harm, no foul, right? Wrong.

The Rising Tide of Sweet Stickiness

Sugar is sticky once it's in your body, akin to the stickiness of maple syrup on your fingers—with the important difference that once sugar sticks to your insides, it can't be washed off. On a molecular level, this is called *glycation*, and it occurs when a glucose molecule bonds to a nearby protein or the surface of a cell, thereby causing damage. Proteins are required for the proper structure and function of all organs and tissues in your body—from your liver to your skin to your brain. Any food that elevates blood sugar has the potential to increase glycation, and any protein exposed to glucose is vulnerable.[8]

FAQ: Should I eat brown rice or white rice?

A: The "healthiness" of grains typically has been assessed with a metric called the *glycemic index*. This is a measure of how quickly the food will affect your blood sugar, but is a minimally useful measure of food quality since it doesn't reflect a typical serving size. Also, when sugars and starches are mixed with other foods, the glycemic index becomes inaccurate, as fat, protein, and fiber delay the absorption of sugar into the bloodstream. A mixed meal of carbs, protein, and fat might actually be even more difficult for your body to deal with than sugar in its isolated form, by prolonging the elevation of insulin. This, over time, can lead to big problems (more on this in the next chapter).

Total glycemic **load**, which takes into account serving size, may be a better measure of meal quality than the glycemic index of any given food. (More difficult to measure but possibly even better would be total **insulin load**, which takes into account the potentiation of fat storage that comes with carbs **plus** fat in processed foods.) Needless to say, stick to the carbs that occur naturally in high-fiber foods like vegetables, low-sugar fruits (I'll list a few in the coming pages), tubers, and beans and legumes, which have a low glycemic index **and** load.

When it comes to rice, just pick the one you prefer. While brown rice contains more fiber and micronutrients than white rice, it's not a great source of either and can be difficult to digest for some people. Given the virtually identical glycemic indexes **and** loads, when we occasionally hit the sushi bar after an intense exercise session, I go for brown rice, while Dr. Paul opts for white (and insists he's made the better-tasting choice for an hour afterward).

Now that you know how easily starches can be converted to sugar, you should be aware that whether you consume a cup of juice, which leads to a sharp spike in blood sugar, or a bowl of brown rice, which contains fiber and sugars linked together in long chemical chains leading to a smaller but prolonged flood, the amount of glycation that occurs for a given amount of carbs is pretty much the same. This rate can be boiled down to a simple formula:

Glycation = Glucose Exposure × Time

Like oxidation, some degree of glycation is going to occur as an inevitable part of life. But the good news is, just as we can slow the rate of oxidation in our bodies by avoiding oxidized oils (among other things), we can also slow the rate at which glycation occurs. And our

Sugar-Rich	Sugar-Poor
Wheat (whole and white)	Grass-fed beef
Oats	Almonds
Potatoes	Avocado
Corn	Fatty fish
Rice (brown and white)	Poultry
Soft drinks	Kale
Cereal	Spinach
Fruit juice	Eggs

most powerful anti-glycation weapon may just be our forks,* which we can use to select foods that do not contain an overabundance of sugar (chained together or otherwise) that can stick to our proteins.

One of the most damaging aspects of glycation is that it leads to the formation of what are called *advanced glycation end-products*, or AGEs—a very appropriate acronym. AGEs are known as *gerontotoxins*, or aging toxins (from the Greek *geros*, meaning "old age"), and are highly reactive, like biological thugs. They are strongly associated with inflammation and oxidative stress in the body, and they are generated in all people, across all ages, and to varying degrees—but dictated by and large by diet.[9] Because AGE formation is more or less proportional to blood glucose levels, this process is dramatically accelerated in type 2 diabetics, playing a major role in their tendency toward the development or worsening of degenerative diseases like atherosclerosis and Alzheimer's disease.

Speaking of Alzheimer's disease, a brain affected by the disease is

* There is clear person-to-person variation of the impact that a carbohydrate meal (say, a whole-grain bagel) will have on one's blood sugar. Someone with healthy glucose control might eat a baked potato and see their blood sugar return to baseline soon after, with minimal damage done. Conversely, a person with poor glucose control (somebody with insulin resistance, prediabetes, or type 2 diabetes) might see their blood sugar stay elevated for hours afterward. These factors are mediated by a range of factors, including inflammation, sleep, genes, stress, and even time of day.

riddled with these aging toxins, containing three times the amount of AGEs as in a normal brain.[10] (Dutch neurobiologist D. F. Swaab, in his book *We Are Our Brains*, actually described the disease as a premature, accelerated, and severe form of brain aging.) Glycation clearly plays a role in this process, and this explains partly why elevated blood sugar increases the risk of dementia, *even among nondiabetics*.[11] But you don't need to have dementia to suffer the effects of AGEs on your cognition. Non-demented adults free of type 2 diabetes with higher levels of AGEs appeared to show an accelerated loss of cognitive function over time, impaired learning and memory, and reduced expression of genes that promote neuroplasticity and longevity.[12]

To get a sense of the rate at which AGEs are being formed in your body, doctors can use a test typically used to manage diabetes called the hemoglobin A1C, which looks at the amount of sugar stuck to red blood cells. Blood cells are in circulation for an average of four months, facing constant exposure to the varying levels of sugar in your blood before getting sent off to retirement in your spleen. The A1C therefore paints a picture of your average blood sugar over the last three months or so, and may be a powerful marker of risk for cognitive decline or even diminished cognitive performance.

In late 2015, I had the opportunity to visit Charité Hospital in Berlin, one of Germany's most research-intensive medical institutions and home of a study that examined the relationship between blood sugar and memory function. The study's lead author, Dr. Agnes Flöel, examined 141 people with A1Cs that were within the range of "normal." She found that for every 0.6 percent increase in a subject's hemoglobin A1C (again, a measure of average blood sugar over three months), two fewer words were recalled on a verbal memory test. That these subjects were nondiabetic, nor even prediabetic, is a startling finding. What's more, people with higher A1Cs also had less volume in the hippocampus, which is the brain's precious memory processing center.[13] (Findings published in *Neurology*, the official journal of the American Academy of Neurology, also indicated that higher fasting blood sugar levels within the range of "normal" were more likely to predict a loss of volume in that brain region.)[14]

DOCTOR'S NOTE: THE DRAWBACKS OF THE A1C

The A1C is not a perfect test, but it reaffirms how damaging sugar can be. Research has shown that elevated blood sugar actually shortens the life spans of blood cells, so while a person with normal blood sugar may have blood cells that live for four months, someone with chronically high blood sugar may have blood cells that live for three or less.[15] The more time spent in circulation, the more sugar a blood cell may accumulate. A person with truly healthy blood sugar might therefore have the "false positive" of an elevated A1C, while a diabetic may in actuality have *even higher* blood sugar than the A1C reveals.

In my clinic, I occasionally use a test called *fructosamine*, which measures compounds that result from glycation and is a reflection of blood sugar control over the previous two to three *weeks*, rather than three months. Unaffected by the varying life spans of red blood cells, this test can be useful to analyze a discrepancy with the A1C, i.e., when average blood sugar is changing rapidly due to dietary adjustment.

Unfortunately, the damage caused by glycation is not confined to the brain. Glycation is known to promote aging of the skin, liver, kidneys, heart, and bones.[16] No part of us is immune. In a sense, your A1C (being directly reflective of both insulin release and AGE formation) may even indicate the rate at which you're aging.

The eyes are a window to another example of glycation-caused aging, as they contain neurons and other cell types highly vulnerable to glycation. Cataracts are a clouding of the lens of the eye and are the principal cause of blindness worldwide. Scientists know that cataracts can be created in as little as ninety days in lab animals simply by keeping their blood sugar high, thereby accelerating glycation.[17] This perhaps explains why diabetics, who have increased rates of glycation, have up to a fivefold increased risk of developing cataracts compared to those with normal blood sugar levels.[18]

Not all AGEs are made in the body, however. Some come from our environment. Cigarette smoke, for example, is a rich vehicle for these aging accelerators to enter the body. AGE formation is also a fairly common chemical reaction in food preparation, particularly when high-heat cooking methods are used. While AGE research is still in its infancy, studies have shown that the vast majority of AGEs are created endogenously—in the body—and are a result of diets rich in carbohydrates. In fact, vegetarians have been reported to have more circulating AGEs than people who eat meat, and it's thought that the reason for this is their higher reliance on dietary carbohydrates and a greater intake of fruit.[19]

AGING TOXINS IN OUR ENVIRONMENT

If you've ever seared a steak on a grill and watched a brown crust develop, you've seen glycation at work. Browning indicates the formation of exogenous (formed outside the body) AGEs. This is known as the Maillard reaction. In truth, the processing of food of any kind creates AGEs, but dry, high-heat methods of cooking like barbecuing or roasting are particularly promotive of AGE formation, and processed meats (sausages and hot dogs, for example) contain a higher amount than their more natural forms. The safest cooking style involves moist heat, such as sautéing or steaming. (Plants will contain fewer AGEs than meat, regardless of cooking style.)

This may cause some to wonder if they should skip the meat altogether, but judging the health quality of a food strictly by its AGE content would be a mistake. Broiled wild salmon, for example, contains a considerable amount of AGEs, and yet wild fish consumption has been associated with healthy cognitive and cardiovascular aging in many studies and trials. Additionally, many anthropologists believe that it wasn't just the consumption of meat but the very act of cooking it that helped our ancestors to extract more calories and nutrients from their food, allowing our brains to reach their robust modern size. The safest way to integrate meat products into your diet is to consume cuts that are organic and grass-fed (or wild, if we're talking fish), which will ensure higher amounts of antioxidants, and to use as little heat as possible (though, of course, you'll need to cook food thoroughly enough to avoid illness).

It's also important to bear in mind that only between 10 and 30 percent of exogenous AGEs get absorbed into your body. Antioxidant nutrients like polyphenols and fiber, which are abundant in plant foods, can also neutralize these aging toxins before they're able to make it into your system.[20] If you'd like to treat yourself to roasted chicken (a fairly rich source of AGEs), for example, opting for a heaping plate of dark leafy greens on the side might help minimize the impact. This also helps broker a more pleasant interaction between those AGEs and the trillions of bacteria that reside in your gut—important players in your brain function, as you'll see.

Added Sugar: The Brain Bane

Added sugar has become one of the worst evils in our modern food supply. Intended by nature to be consumed in small amounts via whole fruit, where it's packaged with fiber, water, and nutrients, sugar has become the pervasive addition to countless packaged foods and sweetened beverages. Now, finally, nutrition labels in the United States are mandated to list the amount of sugar added to products—definitely not a cure-all, but a move in the right direction. Whether the sugar is single-origin organic cane sugar, brown rice syrup, or the industry darling high-fructose corn syrup (HFCS), one thing is clear: the safest level of added sugar consumption is *zero.*

One of the dangers of sugar consumption is that it can hijack our brain's pleasure centers. Packaged foods with added sugar usually taste "impossibly delicious" and cause massive spikes of dopamine, a neurotransmitter involved in reward. Unfortunately, the more we consume, the more we require to reach the same threshold of pleasure. Sound familiar? It should: sugar, in the way it stimulates the release of dopamine, resembles drugs of abuse. In fact, in animal models, rats prefer sugar over cocaine—and rats *really* like cocaine.

To borrow a term from Sigmund Freud, rodents are all *id*—meaning they give in to their cravings. They don't have responsibilities (at least in the human sense), and they certainly don't have to worry about looking good in bathing suits. This is why rat studies are an important part of understanding how food—and in particular sugar—affects our behavior. From rats we've learned, for example, that fructose in particular may *promote its own consumption.* When rats were fed the same number of calories from either fructose or glucose, glucose (like potato starch) induced satiety (a feeling of fullness). Fructose, on the other hand, actually provoked more feeding—it somehow made the rats hungrier. The lesson to be inferred is that sugar, and perhaps especially fructose, may actually be causing you to overeat (more on this below).

These insights are crucial, because we tend to feel guilty when we go through an entire bag of chips (or pint of ice cream, or box of cookies). Been there? Me too. What nobody tells us as we peruse the aisles lined with air-pumped bags of bliss is that these foods are literally engineered to create insatiable overconsumption, designed in labs by well-paid food scientists to be hyper-palatable. Salt, sugar, fat, and often wheat flour are combined to maximize pleasure, driving your brain's reward system to an artificial "bliss point" that simulates the addictive properties of controlled substances. Remember the famous slogan "Once you pop, you can't stop"? It's now a truism with scientific backing.

Foods Uniquely Designed to Screw Up Your Brain

- Bagels
- Biscuits
- Cake
- Cereal
- Milk chocolate/white chocolate
- Cookies
- Energy bars
- Crackers
- Doughnuts
- Muffins
- Pastas
- Pastries
- Pies
- Granola bars
- Pizza
- Pretzels
- Waffles
- Pancakes
- White bread
- Milkshakes
- Frozen yogurt
- Ice cream
- Batter
- Gravy
- Jams
- Jellies
- Fries
- Chips
- Granola

Don't Get "Fruct" Over by Fructose

Beelzebub, Satan, Abaddon, Lucifer, the Evil One—like the Devil, sugar takes many forms and goes by many names. Sucrose, dextrose, glucose, maltose, lactose—what's the difference, and why should you care? They can all spike your blood sugar and tamper with the hormones that control appetite and fat storage. However, one in particular has gotten the spotlight recently, and perhaps for good reason, as it has silently crept its way into every crevice of our food environment: fructose.

FAQ: Now that my favorite soft drink is made with real/organic/non-GMO sugar instead of high-fructose corn syrup, that means it's healthier, right?

A: No! Table sugar (organic or not) and high-fructose corn syrup are both roughly 50 percent glucose and 50 percent fructose. Both are pure sugar, and both can lead to the same problems: addiction, fat storage, and accelerated glycation.

Fructose is processed differently than glucose, bypassing your bloodstream and hopping on the express train to your liver. Dr. Lustig calls the unique effect of fructose on our biology "isocaloric, but not isometabolic" (the *iso* prefix signifies "same"). What this means is that though it has an equal number of calories, gram for gram, as other sugars, fructose seems to behave in a rather peculiar manner from the perspective of your metabolism. It doesn't elevate blood sugar and doesn't cause a rise in insulin—at least not initially. Food companies commonly exploit this difference to sell products sweetened with fructose to health-conscious consumers and diabetics.

Once in the liver, fructose induces *lipogenesis*—literally, fat creation. In truth, all carbohydrates when consumed in excess are capable of stimulating lipogenesis, but fructose may be the most efficient at doing so. One short-term study published in the journal *Obesity* demonstrated nearly double the increase in liver fat when healthy humans on high-calorie diets supplemented with fructose compared to glucose (113 percent versus 59 percent, respectively).[21]

After fructose fills your liver to its capacity with fat, it spills over into your bloodstream as triglycerides. Consuming fat also leads to a short post-meal triglyceride spike, but lipogenesis due to fructose consumption can dump more fat into your blood than even the highest-fat-containing meal—following a high-fructose snack, your blood can actually take on the appearance of pink cream for this very reason. This is also why fasting triglyceride levels (a marker used to assess metabolic health and heart disease risk) are nearly universally influenced by carbohydrate consumption, and by fructose in particular.

While fructose has a negligible *immediate* effect on blood sugar, frequent consumption will eventually cause blood sugar to rise because the stress on the liver creates inflammation, impairing the ability of cells to "suck up" glucose from the blood. This may have been an adaptation to help us store more fat when fruit was in season, and yet now explains why sugar consumption dovetails with skyrocketing rates of type 2 diabetes. (Now is probably a good time to question whether fructose-dominant sweeteners like agave syrup—90 percent fructose—are *really* the right choice for health-conscious individuals or diabetics.)

The combined effects of fructose may add up to altered gene expression in the brain. In a study out of UCLA, rats were given the amount of fructose equivalent to drinking a one-liter bottle of soda every day.[22] After six weeks, they began to show typical derangements: they had escalating levels of blood sugar, triglycerides, and insulin, and their cognition began to break down. Compared to mice fed only water, the fructose-drinking mice took double the time to find their way out of a maze. But what surprised the researchers most was that close to one thousand genes in the brains of the fructose-fed rats were altered. These weren't genes for cute pink noses and fuzzy whiskers—they were comparable to genes in humans, having links to Parkinson's disease, depression, bipolar disorder, and others. The degree of gene disruption was so profound, head researcher Fernando Gomez-Pinilla commented in the UCLA release: "Food is like a pharmaceutical compound" in terms of its effect on the brain. But that power also swings in the other direction—the negative impact that fructose had on both cognition and gene expression was attenuated by feeding the rats DHA omega-3 fat.

Avoiding the stresses imposed on the brain by excess sugar consumption may be an empowering leverage point for the 5.3 million Americans suffering from traumatic brain injury. A diet high in fructose impaired the plasticity of rat brains, diminishing their ability to heal after head trauma. Though rats are not people, brain injury is a fairly organic condition that is easily replicable in animals—unlike, say, a complex human disease that rats and mice do not naturally develop.

Human Foie Gras

Fructose, and sugar consumption in general, is a major contributor to nonalcoholic fatty liver disease, or NAFLD. Currently affecting seventy million adults in the United States (30 percent of the population), rates of NAFLD are expected to explode in the coming years unless we can do something about our collective sweet tooth. By the year 2030, it is estimated that 50 percent of the US population will have NAFLD—and insulin resistance, a problem affecting a staggering number of people worldwide, is directly proportional to the severity of the disease. But we're not the only animals experiencing an epidemic of fatty livers.

Similar to humans but on a much larger scale, ducks and geese are able to store a massive surplus of calories in their livers in the form of fat. This is an adaptation that allows them to fly long distances without stopping for food, and it is exploited to create foie gras, a French delicacy that is enjoyed by many around the world.

Foie gras is a well-fattened duck or goose liver and is revered for its rich, buttery texture—something this type of liver is not typically known for. To make it, tubes are inserted into the throats of healthy geese and ducks and they are force-fed grain (usually corn). With the animals eventually consuming far more carbohydrates than they ever would in a natural setting, the livers swell with fat, growing to nearly ten times their normal size. The swelling can be so severe that it impairs blood flow and increases abdominal pressure, hindering the animal's ability to breathe. Sometimes the liver and other organs will even rupture from the stress. Cruel and inhumane, it provides an excellent, if extreme, illustration of exactly what we're doing to ourselves as a consequence of chronic sugar consumption: developing fat-filled livers and creating foie gras *right inside of our own bodies*.

Save for a dinner date with Hannibal Lecter, you are unlikely to be poached for pâté—but harboring a disgruntled liver can nonetheless produce many undesirable consequences, as the organ is tasked with hundreds of important functions in the body. NAFLD has been associated with cognitive deficits, which increase with the severity of the disease. In mice that are overfed to develop NAFLD, brain changes associated with Alzheimer's disease begin to emerge, and mice that already had Alzheimer's-related abnormalities (not a perfect model of human Alzheimer's, but interesting nonetheless) exhibit exacerbated signs of disease and greater inflammation when fed concentrated fructose.[23]

While 70 to 80 percent of obese individuals have NAFLD, so do 10 to 15 percent of normal-weight individuals, "fed" by the ubiquity of sugar and fructose. As you can see, being thin does not make you impervious to the metabolic and cognitive effects of a poor diet.

A Gut-Brain Terrorist

Unsurprisingly, ground zero for many of the problems associated with sugar is the gut. Fructose in particular, whether from processed sugary foods or excessive fruit sugar, impairs its own absorption when it is consumed in large amounts. While this may sound vaguely positive, the excess fructose left to linger in the gut can create many unpleasant symptoms, from bloating and cramps to diarrhea and symptoms of irritable bowel syndrome (IBS). As gnarly as this sounds, high intestinal fructose concentration can also interfere with tryptophan absorption.[24] Tryptophan is an essential amino acid that we must get from our diets, and it is the direct precursor to the neurotransmitter serotonin, which is important for healthy mood and executive function. This may be why fructose malabsorption is associated with symptoms of depression.[25]

The gut lining is the precious matrix across which we absorb nutrients from food. It also helps keep intestinal bacteria in the gut where it belongs. The last thing you'd want to do is poke holes in your gut lining, but this is what concentrated fructose seems to be able to do. The technical term for this is *increased intestinal permeability*, which is when the gut lining allows the leakage of inflammatory bacterial components from the gut into circulation. The seepage of these bacterial components into your blood is a major driver of systemic inflammation, and it can induce symptoms of depression and anxiety as it switches the immune system of your brain and body into high alert (this phenomenon is covered in greater depth in chapter 7).

While high concentrations of fructose from processed foods have been shown to contribute to gut permeability, this does not seem to happen when small amounts of fructose are consumed in whole, fresh fruit. This is thanks in part to the fibrous matrix that the sugars are bound in, water, and other phytonutrients. Whole-fruit consumption is also self-limiting, because the fiber makes it more filling. For example, it would be quite difficult to eat five apples, but it's a breeze to drink the sugar contained in five apples when juiced.

PLAQUE ON YOUR TEETH (AND IN YOUR BRAIN)

A high-sugar diet might not just put plaque on your teeth—it may deposit it in your brain as well. In an attempt to find out whether blood sugar might increase the production of amyloid plaque (a central feature of Alzheimer's disease), researchers attached glucose clamps to mice that were genetically engineered to develop Alzheimer's-like symptoms. Glucose clamps provide ports that allow scientists to increase or decrease blood sugar levels in awake, freely moving animals, to see what effect such tinkering has on their bodies, brains, and behavior. The researchers then measured levels of amyloid plaque's precursor protein in the animals' spinal fluid.

What the researchers found was fascinating: just by temporarily dialing up the amount of sugar in the mice's blood, amyloid production dramatically increased.[26] Young mice that had their blood sugar doubled during a four-hour "challenge" (the equivalent of eating a high-carbohydrate meal by a person with poor glucose control) had a 25 percent increase in amyloid beta production, as measured in their spinal fluid. Older mice were particularly vulnerable, seeing a 40 percent increase from the same blood sugar challenge.

Researchers noted that repeated surges of blood sugar, such as what is common in type 2 diabetes, "could both initiate and accelerate plaque accumulation." They concluded that plaques in the hippocampus "are likely modulated by blood glucose levels." The important distinction, of course, is that what happens in rodent disease models doesn't always happen in humans. Regardless, studies like this are important pieces in the puzzle for discovering why higher glucose levels are strongly associated with increased risk for dementia, even in people without diabetes.[27]

The Sour Truth about Sweet Fruit

Why would a sugar found naturally in fruit be so poorly tolerated by modern humans? It doesn't make intuitive sense, *until* we consider the scarcity and seasonality of fruit until a few short decades ago.

Like the casino of a Las Vegas resort, our modern food complex has lost any sense of time, place, and season. Within a single generation we have gained unprecedented access to sweet fruit. A pineapple from the tropics, berries grown in Mexico, and Medjool dates from Morocco are now flown to our towns and cities so that they may line our supermarket shelves all year long. These fruits are bred to be larger, and contain more sugar, than ever before in history.

We are frequently told that it's okay—beneficial, even—to consume "unlimited" fruit, but looked at through an evolutionary lens, fruit (and particularly today's cultivated, high-sugar versions) may be uniquely adept at tricking our bodies' metabolisms.[28] This is theorized to be an adaptive, temporary quality that helped us pack away fat so that we might survive the winter. In fact, it is thought that our ancestors developed red-green color vision for the *sole purpose* of distinguishing a ripe, red fruit from a green background—an evolutionary testament to the lifesaving value of fruit for a hungry forager. Today, 365 days of high-sugar fruit consumption is readying our bodies for a winter that never seems to come.

What consequences of gorging on grapes and other sweet fruit could there possibly be for our brains? A few large studies have helped to shed some light. In one, higher fruit intake in older, cognitively healthy adults was linked with less volume in the hippocampus.[29] This finding was unusual, since people who eat more fruit usually display the benefits associated with a healthy diet. In this study, however, the researchers isolated various components of the subjects' diets and found that fruit didn't seem to be doing their memory centers any favors. Another study from the Mayo Clinic saw a similar inverse relationship between fruit intake and volume of the cortex, the large outer layer of the brain.[30] Researchers in the latter study noted that excessive consumption of high-sugar fruit (such as figs, dates, mango, banana, and pineapple) may induce metabolic and cognitive derangements on par with processed carbs.

DOCTOR'S NOTE: WHEN YOU *REALLY* NEED TO RESTRICT FRUIT

People have a wide tolerance for carbohydrates, but for diabetics, it's pretty open-and-shut that sugar, even from fruit, needs to be dramatically restricted. I have my diabetic patients consume fruit in half-serving quantities—even a single orange can spike the blood sugar into an unacceptable range for hours after eating it. But all is not without hope! Once insulin sensitivity is restored, exercise has become a habit, and the system has had time to restore energy balance and metabolic flexibility, unprocessed carb sources can be reintroduced.

Fruits, however, *do* contain various important nutrients. Luckily, low-sugar fruits are among the most concentrated sources of these nutrients. Some examples include coconut, avocado, olives, and cacao (no, this does not mean that chocolate is a fruit—but dark chocolate does have a myriad of brain benefits and is one of our Genius Foods). Berries are also great because not only are they low in fructose but they are particularly high in certain antioxidants shown to have a memory-boosting and anti-aging effect. The Nurses' Health Study, a long-running dietary survey of 120,000 female nurses, found that those who ate the most berries had brains that looked 2.5 years younger on scans.[31] In fact, while a recent analysis of the literature found no association between overall fruit intake and reduced dementia risk, berry consumption was the sole exception.[32] Berry nice!

A Call to Action

Every year, billions of dollars are spent to market junk foods to the American people. But more than simply buying ad space in magazines or on TV, these juggernaut companies regularly fund studies to downplay the role of junk food in the public obesity crisis. The *New York Times* recently exposed scientists involved in a leading soda giant's initiative to shift the focus in the global obesity and type 2 diabetes epidemics from diet to laziness and lack of exercise.[33] An executive of the group was quoted as saying:

> Most of the focus in the popular media and in the scientific press is, "Oh they're eating too much, eating too much, eating too much"—blaming fast food, blaming sugary drinks and so on. And there's really virtually no compelling evidence that that, in fact, is the cause.

While exercise is vital to the health of the brain and body, study after study has shown it to be only minimally impactful on weight compared with what people consume. Fitness enthusiasts know that "abs are made in the kitchen," but for many of those who are overweight and obese, a statement like the above only perpetuates the confusion. This sets up a trap for society's most vulnerable, paving the way for cognitive dysfunction and an early death. This is not an exaggeration: for the first time, our eating habits are killing more Americans than our smoking habits.[34] In fact, the latest figures, published in the journal *Circulation*, suggest that nearly two hundred thousand people die each year from diseases driven by sugar-sweetened beverages alone. That is seven times the number of people killed by global terrorism in 2015.[35]

And speaking of smoking, let's look for a second to the historical awareness of the link between cigarettes and lung cancer. It took decades for enough "proof" to show up in the medical literature to convince physicians that cigarettes were a major driver of soaring lung cancer rates, even though the disease had been "very rare" prior to the ubiquity of smoking in the mid-twentieth century. And who can forget the cringe-worthy ads from the 1940s (easily Googleable) featuring doctors blatantly endorsing cigarettes? As recently as the 1960s, two-thirds of all US doctors believed the case against cigarettes hadn't yet been established, despite smoking being recognized as a leading cause of the lung cancer epidemic two decades prior.[36]

Should we wait for the "scientific consensus" to rethink our consumption of something mired in profit—something for which there is not only no human need, but which data indicates is almost certainly causing us harm? Keep your answer in mind as we venture into one of the great cons of our time in the next chapter.

FIELD NOTES

- Sugar-protein bonding (glycation) can occur to any protein in the presence of sugar. All carbohydrates, with the exception of fiber, have the potential to glycate.
- While end-products of glycation, AGEs, may be consumed in food, the vast majority are formed in your body as a result of chronic carbohydrate consumption.
- Isolated fructose creates stress on the liver, which promotes inflammation and insulin resistance.
- Sugar interacts with genes in the brain, reducing neuroplasticity and impairing cognitive function.
- Certain foods are created to be hyper-palatable, leading to insatiable hunger and overconsumption. These foods are essentially

bear traps of inflammation and obesity, and are best avoided completely.

- Industry does not care about your health; don't wait for "scientific consensus" before banishing unnecessary and potentially harmful substances from your diet.

GENIUS FOOD #3

BLUEBERRIES

Of all commonly consumed fruits and vegetables, blueberries are among the highest in antioxidant capacity because of their abundance of compounds called *flavonoids*. Flavonoids are a class of polyphenol compounds that are found in many of the Genius Foods (you may remember oleocanthal, in extra-virgin olive oil, which is a type of phenol).

The most abundant flavonoids in blueberries are anthocyanins, which have been shown to cross the blood-brain barrier, enhancing signaling in parts of the brain that handle memory.[1] Astonishingly, these beneficial anthocyanins accumulate in the brain's hippocampus. My friend Robert Krikorian, the director of the Cognitive Aging Program at the University of Cincinnati Academic Health Center, is one of the foremost researchers looking into the effects of blueberries on memory function in humans. Dr. Krikorian has published research showing a robust benefit to cognitive function with blueberry consumption; in one such example, twelve weeks of blueberry supplementation improved memory function and mood and reduced fasting blood sugar in older adults at risk for dementia.[2]

Observational research is just as compelling. A six-year study of 16,010 older adults found that consumption of blueberries (and strawberries) was linked to delays in cognitive aging by up to 2.5 years.[3] And while a recent review found no association between general fruit intake and dementia risk in humans, berries *were* found to be associated: they protected the brain against cognitive loss.[4]

How to buy and consume: Fresh blueberries are great, but don't be afraid to buy frozen blueberries, which are often much cheaper (and more widely available) than fresh. Always opt for organic. Blueberries are great in smoothies and salads or eaten as a snack.

Pro tip: All berries are likely helpful to the brain, though they vary in terms of the specific beneficial compounds found in each. When you want to mix things up, blackberries, bilberries, raspberries, and strawberries may be used in place of blueberries.

CHAPTER 4

WINTER IS COMING (FOR YOUR BRAIN)

It's easier to fool people than to convince them that they have been fooled.

—UNKNOWN

This is a story about a jilted lover. After reading the last chapter, it may strain credulity for you to learn that I've had a lifelong love affair with carbohydrates. But it's true: if I had to close my eyes and think of life's greatest pleasures, chomping down on a freshly baked red velvet cupcake would be up there. Let's be honest, though: not everything that we love is good for us.

It doesn't take a nutrition degree to realize that confectionary bakery products are usually loaded with sugar and refined white flour. Even the young me knew to stay away from the obvious offenders and gravitate toward "healthy" grains instead, especially nutty, chewy, whole ones. Despite growing up mostly around white flour in the form of bagels, pasta, and black-and-white cookies (a childhood favorite), I sensed from early on that the less processed grains were, the better they would be for me. Not yet a teenager, I became a lobbyist within my own family, advocating for products with the red "heart healthy" logo on them when I'd help my mom shop for groceries. They were earthier and had more "stuff" in them—like bran—which I believed gave them more bang for the nutritional buck. I even had a favorite bread growing up called Health Nut, the name of which reassured me, as if every slice eaten was another brick laid on the road to health for me and my family.

I carried this appreciation for grains with me into my adult life, and as I became more proactive about healthy eating I thought, as many do, "*More grains equal more health.*" My day would often look like this: In the morning, I'd have a big bowl of granola with fat-free milk or a whole-wheat bagel and a piece of fruit. By lunch, I'd usually find myself starving, grabbing a sandwich or wrap (only on whole wheat), or my favorite—a brown rice bowl. Post-lunch "comas" became a common phenomenon, and as a result, I'd need to have a few snacks to keep blood sugar up in between lunch and dinner—usually a cookie or two, a few whole-wheat crackers, or some dried fruit. (I didn't have an understanding of the dynamics of blood sugar that I do now—and that you will soon—but I did notice that carbohydrates tended to alleviate the lethargy I was experiencing.) My dinner would usually involve more brown rice, but on some nights, I'd switch it up and go for a big bowl of whole-wheat pasta. If there was one rule I followed, it was that meals always had to include a grain.

Even though my energy levels and food cravings often felt like a roller coaster throughout the day, I never thought to question my diet. Why would I? I was in the "1 percent" of grain consumers, eating them nearly exclusively in their whole, unadulterated state. But here's the devastating truth: I was misled about the health quality of grains, and you were too.

Origins of a Myth

One of the most well-known dietary patterns shown to confer cardioprotective and neuroprotective effects is the Mediterranean diet, first popularized by renowned fat disparager Ancel Keys (you may remember him from such popular hits as chapter 2). Keys enjoyed vacationing on the Greek island of Crete, a region of exceptionally long-lived people, and he used their diet as the backbone for his studies in human nutrition. If Keys had visited the East, he might instead have singled out the diet of the exceptionally healthy Japanese, rich in fish eggs, fermented soybeans (a dish called *natto*), and kelp noodles. But Greece and Italy were popular destinations at the time; they were closer and warmer and certainly had better wine.

As he saw it, the people of the Mediterranean built their diets around plant foods and seafood—vegetables, legumes, fish, olive oil, grains, and nuts. But people in the Greek islands also love their meat, and fatty cuts of lamb are regularly enjoyed. This was perhaps lost on Keys, who happened to visit Crete during a particularly lean time, stationing himself on the impoverished island just after World War II, and during Lent, when meat consumption was particularly limited.

Nonetheless, Keys's observations became the basis of the "grain-based" Mediterranean dietary pattern, ultimately informing the development of the highly influential Food Pyramid, which advised consumers to eat less fat and load up on grain products—up to eleven servings per day. (The Food Pyramid's successor, the USDA MyPlate, still advises consumers to include grains at every meal.) Food manufacturers didn't object, taking advantage of hefty grain subsidies. But did Keys attribute the health effects of the Mediterranean diet to the wrong food group?

When looking at population data, one might notice that the intake of whole grains is certainly associated with less diabetes, colon cancer, and heart and brain disease. People who consume mostly brown rice, whole-wheat bread, and highfalutin grains like quinoa tend to make better choices elsewhere in their diets.[1] They may eat more wild fish (rich in omega-3s), extra-virgin olive oil, and vegetables and far fewer of the refined carbs and Franken-oils that are characteristic of the Western diet. They also live healthier lifestyles in general and tend to exercise more.[2] But at this bird's-eye view, it is impossible to isolate the health effects of grains in an overall healthy diet. Still, the idea that whole grains improve health has become, for lack of a better term, ingrained. (They have even been grandfathered into modern "spin-offs" of the Mediterranean diet, such as the government-endorsed DASH diet to reduce high blood pressure.)

In this chapter, as we explore the role of an ancestral hormone called insulin in brain function, we'd like you to don your skeptic's hat (like any good scientist) and consider that the Mediterranean diet is healthy not *because* of grains, but in spite of them.

The Problem with "Chronic Carbs"

Grains are often considered healthy because of the small amounts of vitamins and fiber that they contain. But grains in their most commonly consumed form spike blood sugar as effectively as table sugar. This occurs because the starch they contain is simply glucose molecules bound together in chains that begin to come apart as soon as you chew them.

A major energy precursor in the body, glucose is used to fuel our leg muscles as we ascend a flight of stairs, our brains as we crunch for a test, and our immune systems when we are fighting off a cold. But glucose molecules (from a slice of whole-wheat bread, for example) can't just waltz into cells—they need an escort.

Enter: insulin.

Insulin is a hormone released into the bloodstream by our pancreas when it senses that blood sugar has become elevated. Insulin activates receptors on the surfaces of cell membranes, which dutifully roll out the equivalent of a red carpet, welcoming sugar molecules inside where they can be stored or converted to energy.

When we're healthy, muscle, fat, and liver cells require little insulin to respond. But repeated and prolonged stimulation of insulin receptors will over time force the cell to desensitize itself by reducing the number of receptors on the surface. While tolerance in everyday life is a virtue, tolerance to insulin is anything but. Once it occurs, the pancreas has to release more insulin for the same net effect. Meanwhile, blood sugar continues to creep up and stay higher for longer between meals, accelerating the nasty process of sugar-protein bonding, aka glycation.

Tolerance—or *resistance*—to insulin affects a huge number of people. Newsflash: you might be one of them. Around one in two people in the United States have blood sugar control issues, including prediabetes or type 2 diabetes. The former now affects a whopping eighty-six million people in the United States alone. Type 2 diabetes, the most advanced stage of insulin resistance, develops when a flood of insulin is required to accomplish what at one point would have taken a relatively small amount. Eventually the pancreas "poops out," unable to keep up with the demand for ever more insulin, and blood sugar stays high despite maximal insulin release.

But what of the other half of the population who are neither prediabetic nor diabetic? If your blood sugar is normal, you're all good, right? Unfortunately, even among people with normal blood sugar levels, insulin resistance is startlingly common. Thanks to the work of pathologist Joseph R. Kraft, we now know that abnormal blood sugar is actually a *late* marker of chronically elevated insulin. As it turns out, chronically elevated insulin can evade routine clinical markers (such as fasting blood sugar and the A1C test described in the previous chapter) for years—even decades—before detection, all while impairing your memory and setting the stage for future brain problems.[3]

DOCTOR'S NOTE: THE LONG ROAD TO DIABETES

To provide a sense of baseline: the average healthy adult pumps out about 25 units of insulin per day to control blood sugar. Now contrast that to some of my diabetic patients, who are injecting upward of 100 to 150 units of insulin per day, more than *five times* the physiological norm. This means that before their diagnoses, their pancreases were working double- and triple-overtime for many years before their blood sugar started to creep up.

Priorities for a Different Time

Insulin is the body's chief anabolic hormone, which means that it creates an environment in your body favorable to growth and storage. This can be useful in shuttling energy (in the form of sugar) and amino acids into muscle tissue after a twelve-hour day of pulling weeds from a field or carrying water from a distant well—but more often than not, these resources end up on our hips and waistlines.

To your fat cells, elevated insulin often means one thing: "party time!" This was helpful—lifesaving, even—during more austere times. Today, it's causing our bodies to stockpile fat in preparation for a famine that never seems to come. But while being overweight makes it more likely than not that insulin resistance lurks below the surface, chronically elevated insulin is common among people that are thin. This often goes undetected because most people assume that thinness equals metabolic health—a major error. There's even a medical term for normal-weight patients with metabolic syndrome: *metabolically obese, normal-weight* (the nonmedical term is *skinny-fat*). This illustrates an important but often confused point: insulin resistance and obesity are independent conditions. Yes, it is possible to fit into small sizes and still be "obese" on the inside.

One consequence of elevated insulin for the thin and overweight alike is that the release of stored fat for fuel—a process called *lipolysis*—is blocked. How? Insulin acts like a one-way valve on your fat cells. This means that when insulin is elevated, calories can go in, but they can't come out. Your fat cells become a Roach Motel, meant as a way of increasing (and preserving) stored fuels when the body senses that food is abundant.

Imagine the average person, consuming more than 300 grams of carbohydrate per day, mostly from refined sources like toaster pastries, commercial breads, sugar-sweetened beverages, and wheat snacks. For this person, insulin production is constant. This presents a major problem because certain organs have evolved to use (and even prefer) fat for fuel, such as nerve cells of the eye and the heart muscle, but are prevented from doing so.

New research has shown that, contrary to what was previously believed, fat can be used as a source of energy for the photoreceptors of the eye.[4] Published in *Nature Medicine*, researchers demonstrated how starving these cells of fatty acids could drive age-related macular degeneration (AMD), suggesting that AMD may in fact be a form of diabetes of the eye! In light of insulin's role in suppressing fatty acid release, reducing carbohydrate intake (and thus insulin secretion) could provide a meaningful and safe lifestyle modification for a substantial at-risk population.[5] (AMD remains the leading cause of visual impairment in Westerners over the age of fifty.)

Even the brain can use fat for fuel once fat is broken down into chemicals called *ketone bodies*. Ketone bodies, or simply ketones, become elevated with periods of fasting, very-low-carbohydrate diets, and by consuming certain ketone-producing foods. They are also produced during vigorous exercise, once glucose stores have become exhausted. But ketones aren't *just* a fuel—they also act as a signaling molecule, flipping switches in the brain that seem to have a range of beneficial effects. Among them is the ability to boost BDNF, the brain's ultimate neuroprotective protein. However, chronically elevated insulin keeps us metabolically inflexible by blocking ketones from being generated. "The inhibition of lipid (fat) metabolism by high-carbohydrate diets may be the most detrimental aspect of modern diets," opined Sam Henderson, a well-known researcher in the area of ketones and Alzheimer's disease. (You'll learn more about ketones and all of their therapeutic and performance-boosting potential in chapter 6.)

The key to allowing these fatty acids to come out and play is reducing insulin, plain and simple. Italian researcher Cherubino Di Lorenzo (who actually studies the effects of ketones on migraine headaches) perhaps said it best: "You could think of this [fat-mobilizing] process as the body's own biochemical liposuction."

You Age at the Rate You Produce Insulin

Almost any dieter will benefit from an initial period of drastic carbohydrate restriction. In fact, a very-low-carbohydrate diet will, on average, halve the amount of total insulin secreted by the pancreas and increase a person's insulin sensitivity after just one day.[6] While this is no doubt good news for our beer bellies, muffin tops, and saddlebags, it may also be key to slowing down the aging process.

Aside from contributing to adiposity, the scientific word for fatness, chronically elevated insulin is thought to speed up the underlying processes of aging. MIT and Harvard lecturer Josh Mitteldorf made no bones about it in his book *Cracking the Aging Code*: "Every bowl of pasta sends a message to the body to put on body fat and accelerate the aging process." During a caloric surplus, easily enabled with hyper-palatable carbohydrates, the long-term picture fades from view, and cellular repair projects go on hiatus.[7] After all, why go through the effort of repairing old cells when you can just create new ones from the bounty of available energy?

On the other hand, when the body perceives that food is in short supply, gene pathways involved in repairing and restoring become active so that the body will still be healthy come tomorrow when the famine is over. These pathways are like little biological "apps" hard-coded into our genome that become active in a low-insulin environment.

One such longevity pathway is *FOX03*, which, among other things, helps to maintain stem cell pools in the body as we age.[8] Stem cells are cool because they are able to differentiate into many different cell types—including neurons—and help to repair against damage incurred during aging.[9] Some scientists believe that if we could "top off," or at least slow depletion of, our diminishing stem cell pools as we get older, we'd be able to better defend against the ravages of aging and extend our years spent youthful and healthy. Activating *FOX03* may be one of the most readily accessible ways of doing this. In fact, people with one copy of a gene that makes their *FOX03* more active have double the odds of living to be one hundred years old. (Those with two copies have triple the odds!)[10]

The empowering news here is that we can mimic many of these benefits in part by keeping tight reins on our bodies' production of insulin. We can do this by engaging in brief periods of fasting (which I will introduce in chapter 6), avoiding rapidly digesting sugars, and demoting starches (especially processed grains) from mainstays in our diets to occasional indulgences.[11] (Genius Food #9—wild salmon—also contains a compound that stimulates *FOX03*.)

Gumming Up the Works

You may already be familiar with some of the cognitive consequences of regular insulin spikes—I certainly was. The most obvious symptoms may be the lethargy that you feel soon after a high-carbohydrate meal. This happens because the pancreas, the organ that secretes insulin, isn't an instrument of precision; it's more like a blunt tool, meant to help us store fat during times of plenty (when summer fruit has ripened on trees, for example) to ensure that we survive periods of food scarcity (during winters or droughts). It can be particularly sloppy when tasked with carving the "carbage" from our circulation, often dropping blood sugar to the floor and triggering hunger, fatigue, and brain fog. At this point in the day, we often reach for more carbs and sugary snacks, which treat our withdrawal, tricking us into thinking that these foods are our friends.

The problems associated with chronically elevated insulin, however, extend far past lunchtime. Hyperinsulinemia is now being thought of as a "unifying theory" of chronic disease by some researchers, and its impact on the brain is particularly worrisome.[12] This is perhaps best illustrated by the effects of insulin on a mysterious protein that we produce in our brains called *amyloid beta*.

If this sticky protein sounds familiar, it's because for many decades it was thought to cause Alzheimer's disease. When the brains of Alzheimer's patients were examined upon autopsy, they were found to be riddled with plaques composed of clumps of "misfolded" amyloid protein. The idea that removing the plaques could cure Alzheimer's disease formed the basis of the so-called amyloid hypothesis, but so far, experimental drugs that have reduced the plaque haven't been successful at stopping disease progression or improving cognition. With growing suspicion that amyloid plaque may be more a consequence of an underlying dysfunction than the smoking gun (at least initially), scientists have taken a step back and asked: How do we prevent our brains from becoming an amyloid landfill?

When insulin is elevated (due to frequent high-carbohydrate meals or excessive caloric intake), our ability to break down amyloid becomes handicapped. This is partially due to a protein called *insulin-degrading enzyme* (IDE). As the name suggests, IDE breaks down the hormone insulin, but it also has a side job (who doesn't these days?): it's part of the enzymatic cleanup crew that also degrades amyloid beta. If the brain had a never-ending supply of IDE, it would perform both tasks effectively, but unfortunately, IDE supply is limited, and it has a stronger preference for degrading insulin than it does for amyloid. In fact, the presence of even small amounts of insulin completely inhibits IDE's breakdown of amyloid.[13]

Much of the brain's custodial work occurs while we're off in la-la land. Thanks to the newly discovered glymphatic system, your brain essentially becomes a dishwasher while you sleep, whooshing cerebrospinal fluid around and flushing out amyloid protein and other by-products. As I've mentioned, insulin interferes with the body's housekeeping tasks, and that includes the cleanup that takes place while you sleep. One way to optimize this critical brain cleaning is to stop eating two to three hours before bed to reduce circulating insulin.

If you've ever put a dried bowl of day-old oatmeal into a dishwasher only to find the oats stuck like paste on the bowl even after it runs, you understand the importance of a basic chemical concept: solubility. Amyloid is like oatmeal in the brain. For it to be flushed, it requires that the protein stay soluble, so that it can be dissolved in the cerebrospinal fluid that pulsates throughout the brain. And what makes amyloid as insoluble as dried-out oatmeal?

The deleterious effects of rising blood sugar know no bounds. Sugar binds wantonly to nearby proteins, and amyloid beta is no exception. When amyloid is glycated it becomes stickier and less soluble, and thus less easily chopped up and flushed away.[14] This may explain the findings of a 2015 study, published in *Alzheimer's & Dementia*, which showed that the more severe the insulin resistance in the body (signaling chronically elevated blood sugar), the more plaque buildup there was in the brains of cognitively normal subjects.[15] What's even more surprising is this association held up even among nondiabetic people—meaning just slight insulin resistance is enough to increase amyloid deposition.

The importance of well-regulated insulin signaling to properly maintain the brain highlights the critical necessity of balance between being fed and fasted. Our bodies are adapted to carry out important maintenance tasks in each of these states. We would find little argument that modern life tilts the scales far toward the fed state, which appears to increase brain plaque burden while also preventing important fuels like ketones from reaching the brain. And while amyloid hasn't been established as the causative force in dementia, we're betting that you, like Dr. Paul and me, want to do everything you can to make sure there's less of it around to gum up the works.

Diabetes of the Brain

Before my mother's diagnosis, dementia seemed to me a distant, vague concept evoking images of cute nursing home residents whiling away their final days by shuffling through a fluorescently lit pastel pasture, playing bridge and whining about the food. My disbelief when my mom's diagnosis landed in her *fifties* was underscored only by the shock I uncovered in my research soon after—that in reality, the disease process begins as early as thirty years prior to the first symptom (some data suggests even earlier). When the doctor was giving me the bad news about her illness, he may well have been assigning the same fate *to me*. Still, even in the odd chance I'd wind up with whatever mental monstrosity my mom had developed, a thirty-year window isn't exactly cause for immediate concern, right?

Not quite. Long before the onset of disease, the same factors that may lead to dementia can very likely affect the mechanics of your cognition. I've already explained how insulin facilitates the uptake of glucose into muscle, fat, and liver cells. In the brain, insulin is used as a signaling molecule, influencing synaptic plasticity, long-term memory storage, and the workings of neurotransmitters like dopamine and serotonin.[16] It also helps brain cells process glucose, particularly in energy-hungry regions like the hippocampus.

When biochemical signals become too loud, cells protect themselves by reducing the availability of receptors to hear them. In the brain, a reduced ability to "hear" insulin may negatively affect aspects of your cognition including executive function and your ability to store memories, focus, feel a sense of reward, and enjoy a positive mood.

It's no secret in the medical literature that having type 2 diabetes can reduce cognitive function, but other research has shown that even in a nondiabetic person, insulin resistance is associated with worse executive function and declarative memory—which is what most of us think of when we conjure images of a person with a good memory (and we all want to be that person).[17] One study from the Medical University of South Carolina that examined the brainpower of nondiabetic, cognitively "healthy" individuals found that subjects with higher levels of insulin not only had worse cognitive performance at baseline (when the subjects' labs were first drawn) but exhibited greater declines on a follow-up six years later.[18]

How might you measure your sensitivity (or resistance) to insulin and, thus, get a handle on your brain performance? One of the most important numbers to know is your HOMA-IR. HOMA-IR, which stands for homeostatic model assessment for insulin resistance, is a simple way of answering the question *How much insulin does my pancreas need to pump out to keep my fasting blood sugar at its current level?* It can be calculated using two simple tests that your primary care physician can run: your fasting blood sugar, and your fasting insulin. The formula to determine your HOMA-IR is as follows:

Fasting Glucose (Mg/Dl) × Fasting Insulin / 405

While reference values generally state that any figure under 2 is normal, lower is better and an optimal HOMA-IR is under 1. Anything over 2.75 is considered insulin resistant. The research clearly indicates that higher HOMA-IR values are associated with worse cognitive performance in the present as well as in the future.

Insulin resistance is also extraordinarily common in people with Alzheimer's disease: 80 percent of people with the disease have insulin resistance, which may or may not accompany full-blown type 2 diabetes.[19] Observational studies have shown that having type 2 diabetes equates to a two- to fourfold increased risk of developing Alzheimer's disease. All told, 40 percent of Alzheimer's cases may be attributable to hyperinsulinemia alone, and a growing chorus of researchers and clinicians are now referring to Alzheimer's as "type 3 diabetes." To be sure, type 2 diabetes doesn't cause Alzheimer's disease—if it did, everyone with type 2 diabetes would develop Alzheimer's and everyone with Alzheimer's would be diabetic, neither of which is the case. It does, however, seem increasingly clear that the two are inbred cousins of each other.

The takeaway here is that even below the level of diabetes, or even prediabetes, chronically elevated insulin may be wreaking havoc, impairing the performance of your brain while setting the stage for widespread neuronal dysfunction decades in the future.

A BLOOD TEST TO PREDICT ALZHEIMER'S?

One protein involved in insulin signaling is IRS-1, or insulin receptor substrate 1. IRS-1 is believed to be a highly sensitive marker of reduced insulin sensitivity in the brain. Alzheimer's patients tend to have higher levels of the inactive form of this protein (and lower levels of the active form) in their blood, so researchers from the National Institute on Aging wondered whether a simple blood test may be used to spot those at risk for developing Alzheimer's disease prior to the emergence of symptoms. What they found was striking: higher levels of the inactive form of IRS-1 (signifying impaired insulin signaling in the brain) predicted Alzheimer's disease development in patients with 100 percent accuracy.[20] Even more breathtaking, the difference in these blood markers was evident ten years prior to the emergence of symptoms. This suggests that maintaining the brain's insulin sensitivity throughout life may be a major step toward preventing the disease.

How might we accomplish this? Start with the body. Interventions that appear to improve the metabolic health of the body, when begun early enough, appear to delay the onset or worsening of dementia symptoms. And while metabolic health is ultimately influenced by a myriad of factors—sleep, stress, and nutrient deficiencies, to name a few—the low-carb diet has now been validated as both safe and effective for improving overall metabolic health in dozens of randomized control trials.

The Glycemic Lie

If minimizing frequent and extended insulin spikes throughout the day is our goal, we should be thinking in terms of the total amount of concentrated carbohydrate we consume. This includes more overt sources of sugar, including sugar-sweetened beverages, processed foods, syrups, and pastries. But the truth is that even whole-grain carbohydrate sources commonly referenced as "low-glycemic" like brown rice are rapid and dramatic blood sugar boosters, which must then be shuttled from the blood with the help of insulin. You might not like to hear this, but whole-wheat bread, which was a staple for me for many years, has both a higher glycemic index (a measure of blood sugar impact) and glycemic load (which takes into account serving size) than table sugar! While these whole-grain foods are often described as "better for you" than refined carbohydrate versions, a more accurate way of viewing them is "less bad" when consumed chronically.

FAQ: Does this mean I'll never get to eat grains/sweet potatoes/a banana/my favorite carbohydrate again?

A: No. While the foundation of your diet should always be nutrient-dense, low-glucose foods, insulin signaling is incredibly important, and chronically reduced insulin might be as problematic, albeit for different reasons, as chronically elevated insulin. The occasional higher-carb meal can be helpful for optimizing various hormones and enhancing exercise performance. The window after exercise is generally the safest time for carbs (like sweet potatoes or rice) to be consumed. Why post-workout? After a vigorous workout, your muscles actually **pull** sugar from your blood.

We'll explore the idea of post-workout carbs further in chapter 6.

Another problem is that the glycemic index refers to the food ingested *in isolation*—and the impact of a slice of bread, for example, would be very different when eaten alone versus with fat and protein in a sandwich. As far back as 1983, scientists have known that while adding fat to a carbohydrate meal can reduce the glucose spike, it also increases the amount of insulin released.[21] Put simply, fat can cause the pancreas to overrespond, secreting *more* insulin for the same amount of carbs! (In reality, fat only delays the entry of glucose into the blood, but prolongs the elevation of blood sugar.)[22] This makes the advice often given to those looking to lower the blood sugar impact of foods—to add more fat to a carbohydrate meal to lessen the glycemic spike—misguided.

Other metrics, then, are needed to discuss the hormonal and metabolic impacts of carbohydrate ingestion. Two that are currently under study are the glycemic load and insulin AUC (or area under the curve) for a given meal. Glycemic load essentially takes into account how much sugar a typical serving size of a given food will release into your blood, while insulin AUC is the total amount of insulin a food (or meal) will stimulate. The total impact of a meal on your blood sugar (and your liver's ability to dispose of it) may matter more than how high or how quickly the blood glucose level rises after a single food item. Some research even suggests that fast-releasing carbs—especially in the absence of fat—can be dealt with more quickly by the body, with a short, quick insulin spike, rather than by having elevated insulin for several hours after a mixed meal of, say, a baked potato with butter.

WHAT MAKES A GOOD CARB GO BAD?

The low-carb vs. low-fat debate has been raging in the health sphere for the last decade or so. Zealots on both sides claim a monopoly on the truth, but the truth is that both sides often throw out the evidence that doesn't fit their worldview. There are entire populations that thrive on high-carb, low-fat diets (like the Okinawans of Japan), and those that thrive on high-fat, lower-carb diets (like the Masai of Africa). How do we reconcile the two? Is genetic carb tolerance enough to explain it? A good scientific model of our biology should be able to explain why both can be healthy. What we *do* know is that when indigenous populations around the globe are exposed to a "Western" diet, disease soon follows.

So what makes a high-carb diet suddenly toxic? In examining the differences between a "healthy" high-starch diet and a toxic Western diet, we find a few key points to consider:

- Traditional higher-carb diets are still low in sugar.
- Traditional diets include far less "acellular" carbohydrate—sugar and starch that have been removed from the cells they were contained in. Think whole fruit vs. fruit juice, or sprouted bread compared to pulverized and powdered "whole-wheat" bread. In one recent study mice were fed the exact same amounts of the exact same food, just powdered vs. whole. Guess which group of mice gained the most weight? Powdered. Processing food—carbs, fat, whatever—makes it instantly more toxic to your system.

It's difficult to tease out the harmful effect of sugar vs. sugar *and* fat combined together in addictive processed foods. It may be that sugar, consumed in isolation, isn't toxic or even prone to overconsumption, but becomes so in the context of processing. In truth, it is very difficult for your body to turn small amounts of sugar into fat, but when carbohydrate is present in your system, every molecule of fat consumed alongside it is going to be immediately stored until those carbs are completely used up by your cells. To make things worse, the giant insulin spike that ensues makes that fat inaccessible to your body for energy between meals. This is how hunger snowballs and the loss of metabolic flexibility begins (more on this in chapter 6).

Before we move on, it's important to be aware of the myriad of factors *other* than "chronic carbs" that may contribute to reduced sensitivity to insulin, thereby raising insulin levels and impairing blood sugar control. These include sleep deprivation, genes, exposure to toxic industrial chemicals, and inflammation driven by the consumption of polyunsaturated oils. Research shows that a healthy person who undergoes one single night of sleep deprivation will have impaired insulin sensitivity the following day—essentially making them temporarily prediabetic, all before any carbs are ever eaten!

Chronic stress is another bad guy, able to throw your insulin system out of whack. It can be contributed to by many factors, some obvious, and some not so obvious. Even something as innocuous as noise pollution becomes a major problem in the developed world and can drive chronic, low-grade stress, which can in turn affect metabolic health. A Danish study found that for every ten-decibel increase in traffic noise in proximity to one's residence, there was an 8 percent increased risk of diabetes.[23] This association shot up to 11 percent over a five-year period. We will revisit sleep and stress in chapter 9.

GLUTEN AND YOUR METABOLISM—FRIENDS OR FOES?

Gluten is the sticky protein found in wheat, barley, and rye. Already present in most breads, cakes, pasta, pizza, and beer, gluten is also added to a wide variety of other products for its gooey, mouth-pleasing abilities—but mouths may be the only thing that gluten is making happy. Recent research suggests that gluten may present a unique inflammatory challenge, impairing insulin sensitivity and predisposing one to weight gain irrespective of the carbohydrates with which it is bundled. Case in point: mice that were fed diets with added gluten gained more weight than mice fed the same diets without gluten.[24] These mice had reduced metabolic activity and increased markers of inflammation compared to control mice consuming the exact same number of calories, carbs, and fat—the only difference was they were consuming gluten. The fact that this was a mouse study may tempt you to balk, but don't. "Overall, the mammalian digestive tract is strongly conserved, with major differences between species being likely driven by diet. Given their shared omnivorous nature, humans and mice thus share strong similarities," wrote researchers addressing the question of mouse model utility in studies on the gut in the journal *Disease Models & Mechanisms.*[25] This adds to the growing body of evidence suggesting that gluten's impact extends far beyond the digestive tract—a fact that I will explore in depth in chapter 7.

Making Changes that Last

Making a positive change, such as reducing your grain intake, eliminating sugar, and reaching for nonstarchy vegetables (like kale) over starchy, insulin-stimulating ones (like potatoes) may often seem like a simple act of willpower. However, dietary change is among the hardest things to accomplish for most people. We bring to every meal the accumulation of years of habit, societal pressure, and cultural norms, influencing both what we and our bodies seem to want.

Prior to the obesity epidemic driven by these kinds of foods, people maintained a healthy weight without counting calories or expensive gym memberships. With the following guiding principles, which have worked for both Dr. Paul and me, it will become possible to avoid dense sources of sugar and carbohydrates, and even potentially to achieve weight loss, without counting calories or creating an obsessive relationship with food. (The only time food intake, and thus calories, will be temporarily restricted is during periods of fasting, called *intermittent fasting*, which I will describe in chapter 6.)

Get Your Sleep, and Meditate When Stressed

Stress and sleep loss will sabotage the willpower of an overeater, so it's important to take these into account when considering your diet. We cover both in greater depth in chapter 9, but for now, remember: a good night's sleep allows you the fortitude to make lasting dietary changes by making sure that your hormones aren't working against you.

When it comes to stress, consuming refined grains and sugar can suppress the stress hormone cortisol and stress responses in the brain.[26] This can lead to dysregulation of the body's natural cortisol ebb and flow, and it highlights one of the many addictive pathways stoked by sugar consumption. You should look to reduce cortisol naturally instead, and morning sunlight exposure, meditation, and exercise are just a few easy methods to implement.

Curate the Food Environment

If you are prone to overeating or are a sugar addict, you will probably notice that controlling your food choices is much easier when you are solely in charge. Controlling what you eat at home can be done by tailoring your grocery shopping and stocking your fridge and pantry with whole, healthy, low-carbohydrate foods. Remember: if it's in your shopping cart, it's as good as in your body.

Of course, we can't control every situation. Walking into the office and seeing free cupcakes throws off your carefully constructed environment and likely is when you will most need to use your willpower. It helps to be able to play a mental game, like imagining the food in question as what it is: not food. Or try defusing social pressure with positivity. When being offered junk food by a friend or a colleague that will derail your diet, one strategy is to frame your dismissal in a positive message. Simply saying "I'm good!" and smiling is likely to be more effective than "I'd love to eat that, but I can't" while grimacing. The former sends the message that you are already "whole" and do not need to consume something unhealthy.

The latter reads, "I'm struggling through a diet, but if you give me a pass, I may just budge." (Side note: this trick also works against other forms of social pressure, such as when being offered alcoholic beverages if you'd rather not imbibe.)

When eating out, try looking at the menu in advance and choosing a restaurant that you know will have healthy options. Another pro tip: thank your server in advance for not bringing the bread basket. Who needs *that* staring you in the face? (You can find many more tips for surviving restaurants and supermarkets at http://maxl.ug/restaurantsandsupermarkets.)

Create an Inner "Rule Book" and Write Down Your Goals

I've found that by making healthy living a part of my identity, it's easy to skip the self-negotiation and just defer to my inner rule book. For example, you may decide that you don't eat wheat products, therefore eliminating a nonessential, nutrient-poor food group high in insulin-stimulating carbohydrates. Some other great rules to integrate into your definition of self might be that you'll "only eat red meat if it comes from a kindly treated animal that has been fed what it really wants to eat (that's grass) its entire life," or that you'll "never consume any beverages that have been sweetened with sugar," or that you'll "always buy organic produce when you can afford it." Try writing down your rules and hanging them on your fridge so that you are reminded whenever you reach for a snack. Research suggests that writing down specific goals, called *self-authoring*, significantly increases the chances that such goals become reality.

Forget "Everything in Moderation" and Embrace Consistency

Many people told to "moderate" their carbohydrate intake negotiate themselves down to half a muffin at breakfast and a smaller portion of spaghetti at dinner. While this is less than the Standard American Diet may include, it is still two servings of mainlined glucose (and subsequent insulin secretion) that your body likely didn't need in the first place.

Still, the guidance to eat "everything in moderation" is pervasive. A recent University of Texas study looking to evaluate this unfortunate prescription found that higher dietary diversity, as defined by less similarity among the foods people eat, was linked to lower diet quality and worse metabolic health.[27] Translated to English: participants that abided by the "eat everything in moderation" rule were eating fewer healthy foods, such as vegetables, and more unhealthy foods, such as grain-fed meats, desserts, and soda. "These results suggest that in modern diets, eating 'everything in moderation' is actually worse than eating a smaller number of healthy foods," commented Dariush Mozaffarian, the study's senior author.

"Americans with the healthiest diets actually eat a relatively small range of healthy foods," observed Dr. Mozaffarian. What does this mean for you? Buy the Genius Foods on loop. I'll give you more foods to add to this list in chapter 11.

Have an "Accountabilibuddy" (Real or Digital)

To borrow a term from one of my favorite shows (that's *South Park* for you cord-cutters), it always helps to have an accountabilibuddy, or friend to report to when striving to reach new goals. Send each other pictures of meals, panic texts when close to temptation, and positive encouragement. If you don't have anyone close by to support you, use social media. Alert your friends and followers to your commitment to "regain your brain," and regularly post photos of your meals for encouragement. Create your own hashtag, or feel free to use #GeniusFoods, which I use on my Instagram account (that's @maxlugavere—come say hi!) to highlight meals that incorporate the Genius Foods and "power up" my brain. Your friends want to see you succeed, and you may even inspire them along the way.

A Final Word

Science is always evolving, particularly where the brain is concerned. As I mentioned in the first chapter, 90 percent of what we now know about Alzheimer's disease, the most common form of dementia, has been discovered only in the past fifteen years. The science of dementia prevention (not to mention cognitive optimization) is new; it is certainly not a *settled* science. However, waiting for it to be so could mean years, if not decades, of inaction.

There is a considerable amount of data highlighting how chronically elevated blood sugar (and insulin) may be compromising our cognitive health. And yet the claim that grains (even "healthy" whole ones) *improve* health is made over and over again, with little good evidence to back it up.[28] It's a falsehood that we've become so deeply invested in, it's even represented in the planted landmass of the United States: at least 15 percent is dedicated to wheat, while more than half is dedicated to growing corn and soy. Only 5 percent is dedicated to growing vegetables, which really should be making up half of our plates.

While everyone's carbohydrate tolerance will vary, my recommendation is to fill your plate with foods that are naturally low in carbohydrates and rich in micronutrients and fiber, the latter being a major weapon in our arsenal against chronic inflammation, which I'll describe further in chapter 7. Examples of low-carbohydrate foods include avocado, asparagus, bell peppers, broccoli, Brussels sprouts, cabbage, cauliflower, celery, cucumbers, kale, tomatoes, and zucchini. For your protein and other nutrients, count on foods like wild salmon, eggs, free-range chicken, and grass-fed beef. While my diet in the past was rich in grains, today I make every effort possible to fill my plate with the aforementioned goodies.

Supplying these precious nutrients to your brain is the name of the game in the next section, beginning with a journey into vascular vibrancy. Strap yourself in!

FIELD NOTES

- Reducing frequent, prolonged insulin spikes by minimizing concentrated carbohydrate consumption is one of the top ways of preserving and enhancing insulin sensitivity, thus minimizing inflammation and fat storage.
- Insulin is a one-way valve on your fat cells, preventing the release of stored calories for fuel. Many organs enjoy using fat for fuel, including the brain (once the fat is converted to something called *ketones*).
- Forty percent of Alzheimer's cases may be owed to chronically elevated insulin, which may begin impairing cognitive function decades prior to a diagnosis.
- Grains, including wheat, are dramatic blood sugar and insulin boosters, relatively low in micronutrients, and the top source of calories consumed in the United States. There is no human biological requirement for grains.
- Carbohydrates, albeit important, are but one part of the story—stress, rancid oil consumption, and even toxic industrial chemicals can all contribute to deranged insulin handling.

GENIUS FOOD #4

DARK CHOCOLATE

Did you know that cacao beans were honored as valid currency in the Mexico City region until as recently as 1887? This valuable fruit is as historically revered as it is healthy. It's also among the richest natural food sources of magnesium, according to my friend Tero Isokauppila, a Finnish foraging expert, a medicinal mushroom proprietor, and one of the most knowledgeable guys on cacao I know.

Some of the most impactful benefits of eating chocolate, a naturally fermented food, come from its abundance of flavanols, a type of polyphenol. Cocoa flavanols have been shown to reverse signs of cognitive aging and improve insulin sensitivity, vascular function, and blood flow to the brain, and even athletic performance.[1] Of nearly one thousand cognitively healthy people aged twenty-three to ninety-eight, those who ate chocolate at least once a week were found to have stronger cognitive performance on visual-spatial memory and working memory and tests of abstract reasoning.[2] But how can we make sure the cocoa we're buying is the right kind, given the seemingly endless options at our local supermarkets?

To start, check the label to make sure that the cocoa has not been "processed with alkali," also known as Dutch processing (usually it will say on the ingredients list, right next to the cocoa). Such processing greatly degrades the phytonutrient content of cocoa, taking what would otherwise be beneficial and turning it into empty calories. The amount of sugar contained in store-bought chocolate varies widely—you want something with minimal sugar and a high percentage of cacao, so look for chocolate with a cacao content above

80%. Anything below this and chocolate tends to veer into *hyper-palatable* territory. (Milk chocolate and white chocolate are essentially just candy—pure sugar.) Once you find a good 85% bar, you'll notice that just a piece here and there allows you to enjoy chocolate without creating an insatiable feedback loop that perpetuates until the whole bar is gone.

Better yet—make your own chocolate at home to avoid any sugar at all and eat to your heart's content! It's surprisingly easy—I provide a great recipe on page 335.

How to use: Consume one 85% dark chocolate bar per week. Opt for organic or fair trade certified, which is almost always ethically sourced.

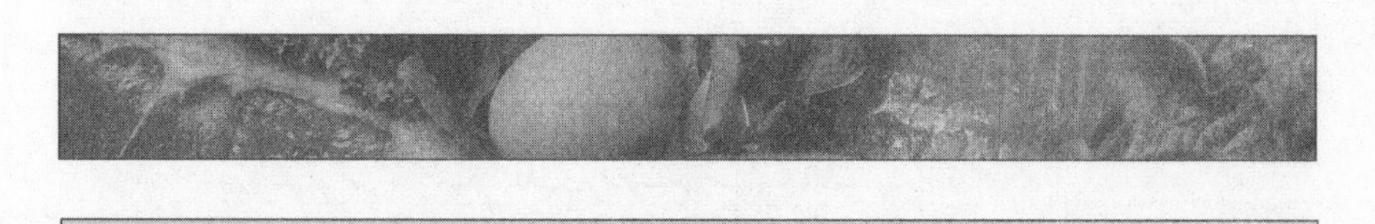

PART 2

The Interconnectedness of It All (Your Brain Responds)

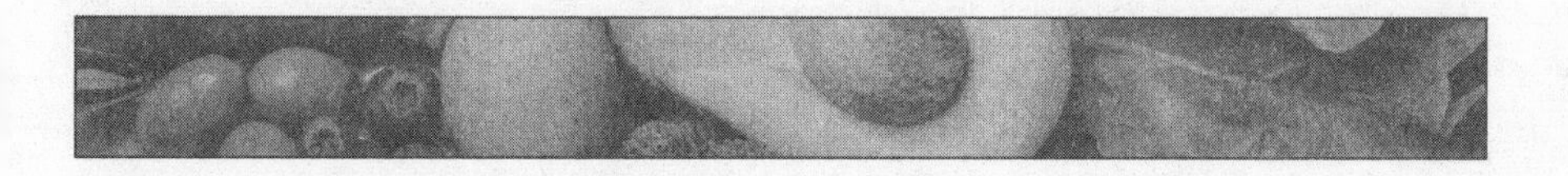

CHAPTER 5

HEALTHY HEART, HEALTHY BRAIN

I remember eating my first omelet like it was yesterday. (What, you don't?) We were in the kitchen of our New York City apartment when my mom whisked up an egg to make an omelet for me. I was seven or eight. My mom was always afraid of heart disease, having had a father who passed from it, which probably explained why growing up I never saw her eating eggs. Their cholesterol-rich yolks, after all, had been in the national doghouse for decades as a causative player in heart disease. One night, however, she offered to cook one for me as a treat.

She calibrated the flame on her beloved cast-iron skillet, given to her by her own mother, which had been seasoned with the corn oil that always sat by the stove.* I sat down at the breakfast bar so that I could watch her and, a couple of moments later, picked up a knife and fork. As she slid the plate over to me, my excitement for what was to be my first-ever egg-eating experience was deflated by a sudden warning from Mom: "You can't eat these too often. The fat and cholesterol in the yolks will clog your little arteries!" (To her credit she also often told me that trying new foods would make me a better lover to the future Mrs. Lugavere. I was always a picky eater, and this was her way of getting me to loosen up. My mom always

* Remember those delicate, chemically reactive polyunsaturated fats from chapter 2? They are great for creating that nonstick coating on your skillet because they oxidize and bond to the iron so easily—the exact process that happens in your blood! It's almost impossible to get a nonstick coating with olive oil or saturated fats because they are more chemically stable and not easily oxidized.

had a quirky sense of humor. Did her promise hold true? Let's just say I remain a picky eater.)

A few years later, we were on vacation in south Florida, which is where many New Yorkers retreat to in order to escape the winter cold. It was there that I had my first taste of another food: coconut. I instantly fell in love with the rich texture, subtle sweetness, and tropical flavor. At the ripe old age of twelve, I understood in that moment why New Yorkers liked Florida so much—the coconuts! But that affair too was cut tragically short when my mom told me that coconut meat was unhealthy. "It's rich in saturated fat, which is bad for the heart."

In this chapter, we're going to take a swan dive into all things vascular health. Why a whole chapter dedicated to blood vessels in a book about the brain? Because the health of your veins and arteries affects more than just the heart and your potential for heart disease. The brain is fed nutrients, energy, and oxygen by a power grid of an estimated four hundred miles of microvasculature. Any outage along this network (leading to reduced blood flow to the brain) not only contributes to cognitive impairment, increasing risk for both Alzheimer's disease and vascular dementia, but can also produce the subtler deficits in cognitive function that we typically associate with aging.[1] And, really, who wants that?

The Diet-Heart Debacle

Today we're armed with a much deeper understanding of vascular health than we'd been in the past, but unfortunately, many doctors still dole out outdated advice. We don't know everything, but it has become increasingly clear that if there *is* a dietary super-villain out there, it is not saturated fat. In 2010, Dr. Ronald Krauss, a top nutrition expert in the United States who was involved in coauthoring many of the early dietary guidelines, concluded in a meta-analysis that "there is no significant evidence for concluding that dietary saturated fat is associated with an increased risk of CHD [coronary heart disease] or CVD [cardiovascular disease]."[2]

Still, the "diet-heart hypothesis"—or the idea that cholesterol in and of itself causes heart disease—persists. This hypothesis originates from initial studies on atherosclerosis, a disease in which plaque builds up to create a hardening and narrowing of the arteries. In these studies, plaques from dissected cadavers were found to be filled with cholesterol. In fact, this is the basis for the catchy, oft-cited idea that "eating fatty foods clogs your arteries," which likens our complex biology to what happens when you pour grease down a cold drain. Because saturated fat does raise cholesterol, and cholesterol-rich foods, you know, have cholesterol in them, reducing intake of both became the focus of efforts for the prevention and treatment of cardiovascular diseases. But biology is seldom simple. As it turns out, cholesterol is often the innocent bystander—present at the scene of the crime, but rarely the villain itself.

Many nutrition scientists, including Ancel Keys, the father of the diet-heart hypothesis, try to reduce whole foods to their constituent "nutrients"—and who could blame them? The discovery of vitamin C cured scurvy. Vitamin D prevents rickets. These were big wins with simple solutions. So when scientists turned their focus to heart disease, it was seductive to buy into the simplistic reductionism: *Cholesterol is found in the arteries of heart attack victims. Eating more saturated fat increases blood cholesterol levels. Therefore saturated fat causes heart disease by increasing blood cholesterol levels.* Just complex enough to be plausible to doctors, and just simple enough to neatly package the story for the public.

But as computer programmers like to say, "Garbage in, garbage out." The enormous complexity and interplay between food and biology often defies our ability to model it, let alone tinker with it by introducing purified or synthetic foodstuffs. Statistician Nassim Taleb, who focuses on randomness, probability, and uncertainty, and who predicted the 2008 financial crisis, makes no bones about it:

> Much of the local research in experimental biology, in spite of its seemingly "scientific" and evidentiary attributes, fails a simple test of mathematical rigor. This means we need to be careful of what conclusions we can and cannot make about what we see, no matter how locally robust it seems. It is impossible, because of the curse of dimensionality, to produce information about a complex system from the reduction of conventional experimental methods in science. Impossible.

In other words, given the incredible complexity of our bodies and our relatively limited scientific tools, we should be intensely skeptical of any rapid, engineered change to our food supply. When the US government stepped in and skimmed the fat off the American diet, our leaders fell into precisely this trap: prematurely applying flawed scientific observations to policy.

Hoping to put the final nail in the coffin for saturated fat, Ancel Keys set up what appeared to be a gold-standard study: a large, long-term, double-blind randomized, controlled trial called the Minnesota Coronary Survey. If you recall from chapter 2, Keys was an epidemiologist—he studied associations in health and disease among large groups of people. This experiment, which involved more than nine thousand institutionalized mental patients, was his chance to prove that the link between saturated fat and heart disease was a causal one, with an ironclad, rock-solid study design.

Keys and colleagues put subjects on one of two diets. The control diet mimicked the Standard American Diet, with 18 percent of calories coming from saturated fat. The "intervention" diet contained only half that—an amount that was in line with the nutritional recommendations made by the American Heart Association and that would later be adopted by the government. To make up for the missing calories, subjects were given foods that were cooked or made with polyunsaturated corn oil, including margarine, salad dressings, and even corn-oil-"filled" beef, milk, and cheeses.

Over five years, the study did show that the corn oil group significantly reduced their cholesterol, but did not show any benefit whatsoever in terms of heart disease or overall mortality.[3] This lack of effect was highly contradictory to much of the nutritional advice being sold to the American people. We were promised that a reduction in blood cholesterol by cutting back on saturated fat intake would *improve* health outcomes, not stall them. This "inconvenient truth" might explain why the study results were published in 1989, a curious sixteen years after the trial's end. But the story doesn't end there.

Max Planck, a Nobel Prize–winning physicist, once remarked that "science advances one funeral at a time," referring to the obstinacy of overbearing and fiercely territorial scientific personalities. This was borne out when, almost thirty years after the initial publication date of the Minnesota Coronary Survey, researchers from the National Institutes of Health and the University of North Carolina discovered unpublished data packed in boxes in the basement of one of the study's now-deceased coauthors—a close colleague of Ancel Keys.[4]

What did the researchers find in this long-buried data? Upon reanalysis, it seemed the corn oil did have an effect on the health of participants, but it wasn't a good one: there was an overall 22 percent *higher* risk of death for every 30 mg/dl drop in serum cholesterol. The corn oil group also had two times as many heart attacks over the five-year period compared to the saturated fat group. Even though the corn oil lowered their cholesterol, they were actually having much worse health outcomes!

The takeaway from this shocking data is that corn and other processed oils (and sugar) are likely much more damaging to your blood vessels than saturated fat. How damaging? Just picture taking a microscopic crème brûlée torch to your arteries, and you'll get the idea. The end result of atherosclerosis looks exactly like fried chicken skin, as physician Cate Shanahan vividly describes in her insightful book *Deep Nutrition*. You'll be dead, but, hey—you'll have lower cholesterol.

Cholesterol and the Brain

It's time for a reality check. Cholesterol is a vital nutrient for the body, and particularly for the brain, where 25 percent of the body's total cholesterol content can be found. It's a critical component of every cell membrane, where it provides structural support, ensures fluidity of nutrient transport into and out of the cell, and may even serve as a protective antioxidant. It's essential for the growth of myelin, the insulating sheath that surrounds your neurons. (Myelin becomes a casualty in multiple sclerosis, an autoimmune condition.) It's also important for maintaining brain plasticity and conducting nerve impulses, especially at the level of the synapse; depletion of cholesterol at this level leads to synaptic and dendritic spine degeneration.[5] Dendritic spines, the branch-like touch-points that facilitate neuron-to-neuron communication, are thought to be the physical embodiment of memories.

Dr. Yeon-Kyun Shin, an authority on cholesterol and its function in the brain, recently published findings in the journal *Proceedings of the National Academy of Sciences* alerting us to the unintended downstream consequences of using blanket cholesterol-reducing medicines (in this case, the ubiquitous class of drugs called *statins*). In the accompanying press release, he elaborated: "If you deprive cholesterol from the brain, then you directly affect the machinery that triggers the release of neurotransmitters. Neurotransmitters affect the data-processing and memory functions. In other words—how smart you are and how well you remember things."

Large-population studies have validated Dr. Shin's fears. In the lauded Framingham Heart Study, an ongoing multigenerational analysis of heart disease risk in residents of the Massachusetts town, two thousand male and female participants were subjected to rigorous cognitive testing. Researchers found that higher levels of total cholesterol, even above the so-called healthy range, were associated with better scores on cognitive tests involving abstract reasoning, attention and concentration, verbal skills, and executive abilities.[6]

Subjects with lower cholesterol displayed poorer cognitive performance. Another study of 185 elderly people without dementia found that higher levels of both total cholesterol (combining both HDL and LDL) as well as LDL alone (often considered the "bad" cholesterol) were both correlated with better memory performance.[7] There is even some data suggesting that higher cholesterol may be protective against dementia.[8]

A recent study of twenty thousand people found strong evidence that those using cholesterol-lowering medications called statins had increased risk of Parkinson's disease, the second most common neurodegenerative disease, which affects movement. "We know that overall weight of the literature favors that higher cholesterol is associated with beneficial outcomes in Parkinson's disease, so it's possible that statins take away that protection by treating the high cholesterol," senior study author and vice chair for research at Penn State College of Medicine Xuemei Huang said in an interview with the website Medscape. (We'll return to statins later in this chapter.)

FAQ: If cholesterol is so good for the brain, I should eat more of it, right?
A: Feel free to enjoy foods that contain cholesterol; just know that there is no need to chase it as a nutrient. This is because the brain naturally produces all the cholesterol it needs. It's more important to ensure the body's cholesterol system stays healthy, and to avoid (as best as possible) medications such as certain statins that might interfere with this synthesis. More on this momentarily.

Cholesterol's duties below the neck affect the brain in important ways as well. It is required to create bile acids, which are essential for the absorption of brain-building fats and protective fat-soluble nutrients. We use cholesterol for the synthesis of many brain-protective hormones such as testosterone, estrogen, progesterone, and cortisol. In concert with UVB exposure from the sun, cholesterol forms another hormone, vitamin D, which is involved in the expression of nearly one thousand genes in the body, many of which are directly involved in healthy brain function.

So at this point, you may be thinking: where do I find this cholesterol stuff? *I want all the cholesterol!* Could we really have been so negligent to have wrongfully accused a nutrient that does so much for us?

The Cholesterol-Disease Connection

Many animal-based foods contain cholesterol, and for many years we were warned that we should limit consuming this fatty substance. Yet the foods we'd worried about for so long, such as egg yolks, shrimp, and other shellfish, actually have a negligible effect on our circulating levels of cholesterol. This is because the body creates cholesterol in amounts much larger than what is found in food. Just to give you a sense, the average person will create the equivalent of four egg yolks' worth of cholesterol in their bodies daily!

DOCTOR'S NOTE: CHOLESTEROL HYPER-ABSORBERS

If we were to write applicable disclaimers before every recommendation we've made in this book, it would become completely unreadable—just keep in mind that we try to frame our words to be applicable to *most people, most of the time.* We posit that dietary cholesterol intake, on the whole, has minimal impact on blood cholesterol levels. As a dietary villain, it's been exonerated, pure and simple. But, and there's always a but, there are specific individuals and genetic variants who are wired differently from most. Most of us synthesize our own cholesterol—but a few people *do* absorb more cholesterol from food! In specific and special cases, especially when managing inexplicably high cholesterol markers surrounding a cardiac event, we can measure blood markers for people with very high internal cholesterol production, or abnormally high cholesterol absorption from food. This can guide therapy when considering why a statin, which blocks cholesterol production, may not be working to lower blood cholesterol levels in a given patient—that person may be absorbing cholesterol from food instead! The specific tests are beyond the scope of this book, but for you citizen-scientists out there, those with elevated *lathosterol* tend to be overproducers who respond better to statins, whereas elevations in *campesterol* and *beta-sitosterol*, plant sterols, indicate overabsorption from the diet.

And yet there's a nontrivial percentage of the population still being told to substitute nutritious egg yolks for alternatives like sugary cereals, instant oatmeal, or worse—the dreadful egg white omelet! A recent Credit Suisse survey exploring consumer perceptions around fat found that 40 percent of nutritionists and 70 percent of general practitioners still believe that eating cholesterol-rich foods is bad for the heart.[9] The authors of the survey wrote:

> The big concern regarding eating cholesterol-rich foods (e.g. eggs) is completely without foundation. There is basically no link between the cholesterol we eat and the level of cholesterol in our blood. This was already known thirty years ago and has been confirmed time and time again. Eating cholesterol-rich foods has no negative effect on health in general or on risk of cardiovascular diseases [CVDs], in particular.

Dietary cholesterol is not, and was never, the problem for most people. Now, even our own FDA has removed cholesterol from the list of "nutrients of concern" in its latest issue of Dietary Guidelines for Americans, putting the final nail in the coffin of one of the most pervasive dietary myths of our time.

As I've mentioned, the vast majority of circulating cholesterol is made in the body, where some of it is produced in our brains, but most of it by our livers. In fact, by eating less cholesterol, we're sending a signal to our livers to create more of it. This is a phenomenon first described by early diet-heart hypothesis detractor Dr. Pete Ahrens, decades ago. On the other hand, the cholesterol that we create in our bodies *can* be related to disease, if we don't keep that cholesterol healthy.

When we create cholesterol in our livers, most of it gets shuttled throughout the body in buses. These buses are your LDL particles; *LDL* stands for *low-density lipoprotein*. LDL is often called "bad cholesterol," but these particles are actually not cholesterol molecules at all, and they're certainly not bad, at least when they're first shipped out. Instead, they are protein-based *carriers* that are essential to helping fat-soluble particles like cholesterol and triglycerides dissolve, or become solvent, in blood. As you likely know, oil and water don't mix, and blood is 92 percent water by volume. In other words, lipoproteins are nature's solution to the solvency problem.

What I've described is a very rudimentary model to understand how cholesterol is made in the body. To begin to understand the link between LDL and disease, you might picture two highways: Highway A and Highway B. The highways have one hundred people on each of them, all commuting to work. On Highway A, those one hundred people are in one hundred different cars. The one hundred people on Highway B are carpooling in five buses. Highway A will be more prone to accidents, pile-ups, and standstill traffic—there are one hundred vehicles on it, after all. Highway B only has those five vehicles—the buses. Which highway would you rather take to work? Unless you're a masochist, sadist, or both, I'm guessing Highway B.

INTERPRETING YOUR NUMBERS

Your typical cholesterol test is akin to estimating road conditions by weighing all of the vehicles on the road, but one bus might have the same weight as five cars, and the standard panel can't differentiate between the two scenarios. The good news is that we now have a test to measure the *total number of vehicles on the road*, and we consider it an invaluable tool. The bad news is most doctors don't know about it, and not all insurances will cover it.

LDL particle number, or LDL-p, can be attained with a test called an NMR (nuclear magnetic resonance) lipid profile. LDL-p represents the total number of LDL *particles*, or vehicles on the road in our highway analogy, which research suggests is a better predictor of risk. And as with our highway analogy, in the case of LDL-p, all else being equal, a lower number is better.

As I've mentioned, cholesterol passengers all start out in buses, as in the Highway B example. These buses are LDL particles that are "large and fluffy" thanks to their many passengers. As the particles drop off passengers, though, they shrink to act more like cars, making them "small and dense." Now, in a healthy system, these smaller particles would return to the liver for recycling before long. However, this process can be disrupted by two maladaptive scenarios—thereby leading to a bloodstream full of small and dense particles. When this happens, your bloodstream looks more like Highway A, which is a sign that your body has a recycling problem.

In the first maladaptive scenario, LDL particles can be damaged, due to either oxidation (a function of time spent in the bloodstream and exposure to oxidative by-products) or the bonding of sugar molecules (this is glycation at work, covered in chapter 3). Once these particles undergo damage, both the target tissues for delivery (your fat or muscle cells, for example) and the recycling center in the liver have trouble recognizing the particles. It's like trying to open a lock with a bent key—the LDL no longer fits. This damaged LDL then gets stuck in circulation and accumulates like a roaming leper colony, eventually settling in an artery wall. Sometimes this means that total cholesterol will go up, but if the particles are small and dense, total cholesterol may not go up much, if at all. This may explain why many people who've never had high cholesterol (or people on medication with artificially lowered cholesterol) still have heart attacks.

The second scenario is that the lock itself may become jammed. This occurs when the liver undergoes oxidative stress and overload, due to excessive processed or concentrated carbohydrate consumption (among other things). Essentially, when the liver is digesting carbs (or carbs and fat simultaneously), alcohol, or other toxins, it doesn't prioritize the recycling of lipoproteins. Similarly, when a target like a muscle cell is already "topped off" with nutrients, it's going to say "no thank you" as the LDL particle passes by. Either situation results in additional time that an LDL particle spends in circulation and in proximity to oxidative by-products—thereby facilitating damage and making it more likely to get stuck to the vessel wall. (This was demonstrated in a recent study where women who went on a high-carb, low-fat diet saw their levels of oxidized cholesterol surge by 27 percent, even though total cholesterol did not change.)[10]

LDL-RECYCLING "CHEAT CODES"

Easing the processing load on the liver may lead to a healthier lipid profile, particularly in certain genetic populations that have a different response to ultra-high-fat or high-saturated-fat diets. The common gene variant associated with increased risk for Alzheimer's disease, *ApoE4*, is thought by some to promote an exaggerated blood-lipid response to saturated fat—namely, increased LDL—in the 25 percent of the population who carry it.[11] Though the mechanism is not fully understood, some researchers suspect that it is owed to reduced LDL recycling by the liver, which can cause LDL to spend more time in circulation, growing smaller and thereby causing problems. These tactics are likely to help make your liver an LDL-recycling superstar:

- **Regain insulin sensitivity.** Eliminate processed grains (even whole wheat), inflammatory oils, and added sugars (especially fruit juice, agave syrup, and high-fructose corn syrup), and reduce consumption of sweet fruit and starchy vegetables.
- **Consume more extra-virgin olive oil.** A diet rich in monounsaturated fats (compared to a high "healthy" carbohydrate diet) reduced liver fat 4.5-fold in a study of diabetics who had excess fat in their livers. Avocados and avocado oil, macadamia nuts, and extra-virgin olive oil are great sources of monounsaturated fats.
- **Reduce consumption of "added" saturated fats.** Saturated fat reduces LDL receptors on the liver, raising LDL.[12] Avoid excessive butter, ghee, and coconut oil. Whole-food sources (such as grass-fed beef) are fine.
- **Load up on fibrous veggies.** This can slow the absorption of both carbs and fat, giving the liver more time to process a meal.
- **Reduce or eliminate alcohol consumption.** A six-pack of beer can cause instantaneous fatty liver in healthy young men—in the course of one sitting!

- **Integrate periods of intermittent fasting, which boosts LDL recycling.** More on fasting in the next chapter.
- **Integrate higher-carb, low-fat post-workout meals one to two times per week.** Once insulin sensitivity is regained, insulin can be used to "turn on" the liver's LDL recycling machinery. Sweet potatoes or white or brown rice are good low-fructose options to help kick that process into gear.

Once one of these now-toxic LDL particles penetrates the vessel wall, adhesion molecules are released to mark the site of injury. Then, various pro-inflammatory messengers called *cytokines* are secreted, which alert your immune system to a breach. This promotes the accumulation of immune cells that stick themselves to the site of the action, forming what is called a *foam cell.* When multiple foam cells coalesce, they create a characteristic fatty streak, marking the beginning of what, over time, might become a plaque, as other immune cells, platelets, and dysfunction of the artery wall compounds.

The process of LDL oxidation clearly plays a major role in the development of atherosclerosis. Interestingly, atherosclerosis is only found in arteries, as opposed to veins. Arteries, unlike veins, carry oxygenated blood in a high-pressure environment, providing fertile ground for those small, dense LDL particles to become damaged and stick to the vessel wall. And while a heart attack (due to plaque build-up in the arteries surrounding the heart) is what many would consider a worst-case scenario, atherosclerosis can happen anywhere, including the microvasculature supplying oxygen to the brain. This is what vascular dementia is: lots and lots of tiny little strokes in the brain.[13] And it is the second most common form of dementia after Alzheimer's.

But what if you're young and healthy, decades away from that brain disease "only old people get"? Can this elegant plumbing system really affect your cognitive function? My friend and colleague Dr. Richard Isaacson, who heads up the Alzheimer's Prevention Clinic at Weill Cornell Medicine and NewYork-Presbyterian, has seen countless patients whose elevated levels of small, dense LDL particles have correlated with lower-than-expected executive function on cognitive tests (this includes the ability to think clearly, focus, and be mentally flexible). While the exact mechanism is unclear, it's plausible that the underlying processes described above are contributing in some way. Dr. Isaacson is now rigorously studying these associations in an effort to validate his clinical observations.

INCREASING BLOOD FLOW TO THE BRAIN

Our brains are massive oxygen consumers. Twenty-five percent of every breath you take is going directly to support the ravenous metabolic needs of your brain, and ensuring that your blood lipids are healthy is one way of keeping your cognitive power supply free of interruption. Thankfully, there are some other ways of increasing healthy blood flow to the brain:

- **Eat dark chocolate.** Compounds in dark chocolate (called *polyphenols*) have been shown to boost brain perfusion, or blood flow to the brain. As we learned with Genius Food #4, stick to 80% or higher for the cocoa content (ideally 85% or higher—this means less sugar), and make sure the chocolate has not been processed with alkali, which degrades antioxidant content.
- **Eliminate or reduce grains, sugar, and starch.** Allowing your brain to run on fat, or more specifically *ketones*, may increase blood flow to the brain by *as much as 39 percent*.[14] More on this in the next chapter.
- **Consume more potassium.** High-potassium foods include avocado (a whole avocado has twice the potassium of a banana!),

spinach, kale, beet greens, Swiss chard, mushrooms, and, believe it or not, salmon.

- **Indulge in nitrate-rich foods.** Nitric oxide dilates blood vessels and expands your arteries while also improving blood flow. Gram for gram, arugula has more nitrates than any other vegetable. Close seconds include beets, butter leaf lettuce, spinach, beet greens, broccoli, and Swiss chard. A single nitrate-rich meal may boost cognitive function.[15]

Can Heart Disease Begin in the Gut?

One last and underappreciated means by which the small, dense LDL particles may overexpress in the body is through an unhealthy gut.[16] Within the sanctuary of our intestines resides an impressive population of bacteria. Most of the time, these bacteria are friendly and enhance our lives in invisible ways. But when we neglect to maintain their home turf, bacterial fragments can "bleed" into our circulation, causing major problems.

One of these normal bacterial components is lipopolysaccharide, or LPS, also known as bacterial endotoxin (meaning "internal toxin"). Under normal circumstances, this endotoxin is kept safely in your intestines, much like the highly corrosive hydrochloric acid that is kept in your stomach. But unlike the stomach, the lower GI tract is a place of active transport of nutrients into circulation. It's quite a selective system, but as a result of our Western diets and lifestyles, the barrier that controls these transactions can become inappropriately porous, allowing LPS to seep through.

One way our bodies may provide a means of damage control is by sending LDL cholesterol carriers to the rescue, like firemen tasked with putting out a fire. It's believed that LDL particles serve an antimicrobial purpose, containing docking sites called *LPS-binding proteins* that allow them to soak up renegade LPS.[17] When the liver senses that LPS has entered circulation via inflammation signaling, it ramps up production of LDL to bind and neutralize it. A chronically "leaky" gut may therefore drive LDL through the roof. On top of that, once the LDL binds with LPS, the endotoxin may affect the liver's ability to dispose of these toxin-carrying particles, creating a double whammy of trouble. A small but increasing number of cardiologists believe heart disease to originate in the gut for this very reason.[18]

Here are just some of the ways you can protect your gut to promote healthy LDL levels:

- **Consume lots of fiber.** Dark leafy greens like spinach and kale are excellent sources of fiber, along with asparagus, sunchokes, and alliums like garlic, onions, leeks, and shallots. Start slow and work your way up to avoid any digestive discomfort.
- **Double down on raw probiotic-containing foods.** Kimchi, sauerkraut, and kombucha—my personal favorite—are great options.
- **Get in lots of polyphenols.** These benefit both you and your gut microbes directly. Good sources include extra-virgin olive oil, coffee, dark chocolate, and berries. Onions are also great to support gut barrier function.
- **Cut sugar from your diet, especially in the form of added fructose.** Fructose, whether from organic table sugar (sucrose is 50 percent fructose, 50 percent glucose), agave syrup (90 percent fructose), or high-fructose corn syrup (which is actually 55 percent fructose), not only increases intestinal permeability but facilitates the leakage of LPS into circulation.[19] Low-sugar whole fruits are fine, because they come packaged with fiber and nutrients that support the gut's own resistance to permeability. Vive la résistance!
- **Eliminate wheat and processed foods from the diet.** Gluten (the protein found in wheat and added to a multiplicity of pro-

cessed foods) has the potential to expand the "pores" in the gut lining. This effect may be magnified by low-fiber diets and additives common in processed foods. This topic is explored further in chapter 7.

DOCTOR'S NOTE: HDL—THE GOOD, THE BAD, AND THE FUGLY

Before medical school, when a doctor started talking to me about "good cholesterol" and "bad cholesterol," my eyes glazed over. Say what? Now, when a doctor talks to me about "good cholesterol" and "bad cholesterol," my eyes *still* glaze over, but for a different reason—because when you go down the rabbit hole, the good/bad analogy becomes laughably simplistic.

The previous section focuses on the LDL story because, all else equal, more LDL particles floating around for longer means higher disease risk. High-density lipoproteins, or HDL, the "good cholesterol," on the other hand, are less well understood—but just like LDL, the total amount of cholesterol in your HDL test may be less important than the number of healthy, functional particles you have.

HDL particles are thought to benefit your health because they are sort of like the cleanup trucks. They pick up excess cholesterol from the far reaches of your body and deliver it back to your liver, where it is converted to bile and passed. In fact, a low HDL-LDL or HDL-triglyceride ratio is a stronger predictor of heart disease risk than high "bad cholesterol." Interestingly, saturated fat—while it does raise the amount of LDL in the body—*also* raises HDL, maintaining a cardiovascular-favorable lipoprotein ratio.

But the amount of HDL you have isn't the whole story. Newer tests are being developed to look at the *functionality* of the HDL recycling system. We call it *efflux capacity*: how efficiently your HDL scavenges cholesterol from the overworked white blood cells in your damaged arterial plaques and shuttles it back to the liver.

The other aspects of functional HDL are still being discovered. It acts as a potent antioxidant and anti-inflammatory, it supports vessel health by promoting the creation of nitric oxide, a gas that keeps your blood vessels dynamic and open, and it may even have an anti-clotting component.

Okay, so you love HDL as much as we do. Now how do you make it more functional?

You guessed it—a lower-carb diet. Adults with metabolic syndrome (more than one in two US adults) generally have low HDL, high triglycerides, and elevated blood pressure, blood sugar, and abdominal fat. A lower-carb, high-fiber diet reverses all of these factors, and returns you to a state of metabolic health. When you consider that even mildly elevated blood sugar *alone* increases your heart attack and stroke risk by 15 percent, this is a no-brainer.

One last thing—HDL proteins are probably no less sensitive to the "biochemical blowtorch" of oxidative stress from rancid polyunsaturated fats and sugar than the LDL proteins are, so you can kill two birds with one stone by reducing your consumption of processed vegetable oils!

Statins: The Brain Drain

One result of the widespread fear surrounding cholesterol is the meteoric increase of prescriptions for a type of cholesterol-lowering medication called statins. If you're still a few decades away from being prescribed one, chances are you'll find a bottle in your parents' medicine cabinet. It's clear what they are because their chemical names all end in *-statin*. An estimated twenty million Americans take statins, making them the most widely prescribed class of drug in the world. The most commonly sold variant is *rosuvastatin*, which regularly makes it to number one on the list of top-selling drugs in the United States. This is a big business, netting pharmaceutical companies $35 billion in sales in 2010.

Long before my mom began to show signs of cognitive decline, she was put on one of these drugs, when it was determined by one of her physicians that her elevated cholesterol needed to be treated. Even though she's never had a heart attack or stroke, when she told me over the phone that she began taking the drug (I was in Los Angeles at the time), I assumed that it was safe and par for the course of "getting older." Plus, a doctor prescribed it. How could it be anything *but* safe?

The problem is, statins are not like seat belts—they often have unintended side effects, or what my friend psychiatrist Kelly Brogan calls, simply, "effects."

As you've learned, cholesterol is important for many things, including immunity, hormone synthesis, and healthy brain function. Evidence suggests that while statins do reduce total LDL, they do little to reduce the proportion of *small* LDL, which is actually the most risk-promoting, easily oxidized variant of LDL. This is because statins decrease the amount of LDL created by the liver, but don't solve the underlying LDL recycling problem described earlier. In fact, some studies have shown that statins can actually *increase* the proportion of small, dense LDL.[20] Many physicians, however, do not distinguish between LDL patterns before taking out the prescription pad. (To learn the dominant particle size, along with the number of LDL particles in your blood, ask your doctor to run an NMR lipid profile.)

Dr. Yeon-Kyun Shin, whom I mentioned earlier, is among the scientists to validate the notion that cholesterol-lowering drugs can also lower the brain's production of cholesterol. "If you try to lower the cholesterol by taking medicine that is attacking the machinery of cholesterol synthesis in the liver, that medicine goes to the brain too.

And then it reduces the synthesis of cholesterol which is necessary in the brain," he said in an Iowa State University release.

Because the brain is made largely of fat, statins with a higher affinity for fat are more easily able to penetrate the brain. Atorvastatin, lovastatin, and simvastatin are lipophilic, or fat-loving, and can cross the blood-brain barrier more easily. Countless reports have been made of these lipophilic variants inducing cognitive side effects, mimicking dementia in extreme cases.[21] (The drug that my mom was on at the onset of her cognitive symptoms was lovastatin.) On the other hand, pravastatin, rosuvastatin, and fluvastatin are more hydrophilic, or water-loving, variants, and may be somewhat "safer" options.

Statins also lower levels of coenzyme Q_{10} (CoQ_{10}), a nutrient important for brain metabolism. As you're going to learn in the next chapter, brain metabolism is vitally important, and decreased metabolism has been linked as the earliest measurable feature in preclinical Alzheimer's disease. CoQ_{10} is also a fat-soluble antioxidant, and helps to keep oxidative stress in check. Lowering it with statins could be bad news for the oxygen- and polyunsaturated fat–rich brain.[22]

DOCTOR'S NOTE: WHY WE'RE WARY OF STATINS

The paradigm under which statins were originally studied, and for which the strongest data in support of their use exists, was secondary prevention—preventing a heart attack *after* you've already had one. The indication for their use was expanded to primary prevention (preventing a cardiovascular event in someone who's never had one) through drug company–funded studies, essentially labeling millions of Americans who had never had a heart problem as now having the "disease" of *hypercho-*

lesterolemia, or high cholesterol. But that's good, right? We are saving lives! The key concept here, however, is that most of those people put on a statin for high cholesterol *would never have had a heart attack in the first place.* I'll say that again: the largest proportion of people who take statins are healthy individuals. The statin is helping *someone*, but for each person it helps, we as physicians have to give it to potentially hundreds of perfectly healthy people with the accompanying side effects and no health benefit whatsoever.

One way we quantify the overall effectiveness of a drug is the NNT, or "number needed to treat." To illustrate, let's look at car seat belts. If they are worn as a preventative tactic with widespread use, their efficacy is substantial, and the risk of serious side effects is close to zero. Many people have to wear seat belts in order to save one life—a massive NNT. But this is not much of a problem, as wearing a seat belt confers no side effects. This is not the case for statins, which regularly cause muscle pain, memory problems, and metabolic dysfunction, greatly increasing the risk of diabetes and even parkinsonism in otherwise healthy people.

So what's the NNT for statins in at-risk adults with no known cardiac disease? Studies range from 100 to 150 to prevent one cardiac event (heart attack or stroke), with no effect on death rate. In other words, ninety-nine of one hundred subjects would have no benefit from the statin. If this were something with minimal cost and zero side effects like a seat belt, we might be able to justify giving those ninety-nine people this extra medication. But this is where the inverse of NNT comes in—NNH, or "number needed to harm." For statins, the NNH to develop muscle damage (myopathy) is nine, or about one in ten patients, and the NNH for diabetes is 250. There is no right or wrong answer to "Should I be on a statin?" That said, you and your physician should be able to have this informed conversation and come to a decision about what you put into your body and why. Unfortunately, most physicians are too pressed for time in the current insurance climate to have such a nuanced conversation with every patient, which means they are forced to take shortcuts and generally end up overtreating or using cookie-cutter guidelines.

As a primary care physician, I prescribe statins very selectively, generally only in the case of secondary prevention, i.e., after a cardiovascular event—and sometimes not even then. I always partner with my patients to come up with a global risk reduction plan (including many of the recommendations in this book!), of which diet and exercise are the cornerstones.

Another means by which statins can affect the brain, both directly and indirectly, is by nearly doubling your risk of developing type 2 diabetes. Published in 2015, a very large, long-term study involving 3,982 statin users and 21,988 nonusers (all *with the same risk factors* for diabetes) found that though all subjects began the study in normal metabolic health, the statin group had twice the rate of diabetes after ten years, and more ended up overweight.[23] Remember: having type 2 diabetes increases your risk for Alzheimer's disease two- to fourfold, along with any number of other chronic diseases including heart disease.[24]

At this point you may be wondering, if statins are so widely prescribed, are they helping *anybody* aside from Big Pharma's bottom line? For patients who already have cardiovascular disease, statins provide an anti-inflammatory effect, independent of their effect on cholesterol. As I've mentioned, inflammation is a major driver of not only cardiovascular disease but diseases of the brain as well, and for this reason, statins may confer some smidgen of benefit. But why put yourself through all of the side effects I've just mentioned when inflammation can be modulated by diet and lifestyle?

What I hope you take away from this section, even if you're not currently on a statin drug, is a sense of how intricately connected the systems of the body are. Though your primary care doctor may prescribe a statin based on a report of "high cholesterol" and send you on your way, drugs *do not work in isolation*. Nor, as you've learned, do compounds manufactured by our own bodies.

So, reduce carbohydrate and polyunsaturated fat consumption—and eat all the coconut and omelets you want—while you let your cholesterol continue to safely fulfill its many important roles in your body. Up next, how to tap into the universe's most advanced hybrid fueling technology—and I'm not talking about a car.

FIELD NOTES

- Cholesterol is critical to an optimally functioning brain and body, but its mode of transportation, the LDL particle, is highly vulnerable to the insults of the Western diet and lifestyle.
- Avoid sugar, refined carbohydrates, and potential gut-busting insults like chronic stress and a fiber-deficient diet, which take a good thing (your healthy LDL particles) and make it bad. Cholesterol, the passenger on the LDL particle, is often simply an innocent bystander.
- Polyunsaturated oils are easily oxidized and torch the inside of your blood vessels.
- LDL damage is a product of poor recycling. Easing the processing load on the liver will help it more effectively recycle LDL, preventing it from forming small, dense particles that can develop into plaque in your arteries.
- Statins are a brain drain—have a conversation with your doctor before you start them or discontinue them for primary prevention (that is, preventing a heart problem if you've never had one).

GENIUS FOOD #5

EGGS

Concerns about the "dangerous" cholesterol content in egg yolks have been debunked. Recent large, long-term studies have elucidated that even a high degree of egg consumption does not increase risk for cardiovascular disease or Alzheimer's disease—in fact, eggs actually boost cognitive function and markers for cardiovascular health. One study, performed in men and women with metabolic syndrome, found that with a reduced-carbohydrate diet, three whole eggs per day reduced insulin resistance, raised HDL, and increased the size of LDL particles to a much greater degree than the equivalent supplementation with egg whites.[1]

In an embryo, the nervous system (which includes the brain) is among the very first systems to develop. Therefore, an egg yolk is perfectly designed by nature to contain everything needed to grow a healthy, optimally performing brain. This helps make eggs, and especially the yolks, one of the most nutritious foods you can consume. They contain a little bit of nearly every vitamin and mineral required by the human body, including vitamin A, vitamin B_{12}, vitamin E, selenium, zinc, and others. They also provide an abundant source of choline, which is important for both healthy, flexible cell membranes and a learning and memory neurotransmitter called *acetylcholine*. And egg yolks contain lutein and zeaxanthin, two carotenoids shown to protect the brain and improve neural processing speed. In one Tufts University study, eating just 1.3 egg yolks per day for 4.5 weeks increased blood levels of zeaxanthin by 114 to 142 percent and lutein by 28 to 50 percent—wow![2]

How to use: Enjoy liberal consumption of whole eggs. Scramble them, poach them, fry them (in butter or coconut oil), or soft-boil them. Since egg yolks contain many valuable fats and cholesterol that are vulnerable to oxidation, I recommend keeping the yolk runny, or more custard-like, as opposed to cooking it through (hard-boiled, for example). For scrambles and omelets, this means using low heat and keeping the eggs creamy or soft as opposed to dry and hard.

How to buy: With so many egg varieties available, it can be confusing to know which ones to buy—and it will often depend on your food budget. Here is a simple metric to help guide your choice:

Pasture-raised > Omega-3-enriched > Free-range > Conventional

Regardless of variety, eggs are always a low-carb, inexpensive, and highly nutritious choice (even conventional eggs, if that's all your budget allows). They are perfect for breakfast, but can be great with any meal—even dinner. And, most important, eat the yolks, folks!

CHAPTER 6

FUELING YOUR BRAIN

We've already covered how diet can help you attain the most receptive possible membranes for your eighty-six billion brain cells. We've discussed how to deliver healthy blood and nutrients to those cells, as well as why we should keep insulin signaling well regulated and blood sugar low. But what we have yet to discuss are the *engines* of those cells, the organelles responsible for keeping the lights on: mitochondria.

Right at this very moment, we are amid a global energy crisis. It's not something you'll read about in the newspaper, it's not the beneficiary of expensive fund-raisers and galas or scientific grants, and it isn't the subject of a dozen Netflix documentaries with A-list actors as executive producers. It may, however, be responsible for mental fatigue, insatiable hunger, brain fog, forgetfulness, and widespread cognitive decay.

Your brain requires a tremendous amount of fuel to function properly. Despite its relatively tiny mass—2 to 3 percent of your body's total volume—it accounts for 20 to 25 percent of your resting metabolic rate. This means that one quarter of the oxygen you breathe and the food you eat is being used to create energy to fuel your brain's many processes. Whether studying for a test, preparing for a speech, or swiping through your favorite dating app, your brain is burning through fuel at the same rate as the leg muscles during a marathon race.[1]

Our energy crisis is not from a shortage of fuel, however. Our brains, if anything, are overfueled. For the first time in history, there are more *overweight* than *underweight* humans walking the Earth.[2] So what, then, accounts for the cognitive malaise?

Punished at the Pump

By the middle of the twentieth century, petroleum-based gasoline became the fuel used by the vast majority of cars on the road. It is only now, decades later, that we realize our gasoline addiction has many long-term side effects and unintended consequences, not appreciated until potentially irreversible havoc has been wrought on the environment and our health.

Glucose, one of the brain's primary forms of fuel, is in many ways like gasoline, and it enters the blood by way of the carbohydrates we consume. A warm sourdough roll? Glucose. A medium baked potato? Glucose. A wedge of highly cultivated sweet pineapple? Glucose (and fructose). When you consume it frequently, glucose provides the main source of energy for the brain. From this sugar, our mitochondria generate energy at the cellular level through a form of complex combustion involving oxygen. This process is called *aerobic metabolism*, and life as we know it would be impossible without it. But as with gasoline, metabolism comes at a cost: exhaust.

One of the by-products of glucose metabolism is the creation of compounds called *reactive oxygen species*, or free radicals. These damaged zombie molecules are the same as those described in chapter 2, and their presence is a normal and unavoidable aspect of living. Right now, as you read this, mitochondria throughout your body and brain are converting glucose and oxygen to energy and leaving behind these waste products as a result.

Free radicals aren't all bad—during exercise, their concentration is momentarily increased and they become powerful signaling mechanisms, coaxing the body to adapt and detoxify in powerful ways. (I will cover this in greater detail in chapter 10.) Under ideal circumstances, we have the ability to clean up these compounds. But when excessive free radical production is sustained, it can outrun our body's ability to effectively mop things up, thereby kicking off a cascade of damaging processes that drive aging and its associated conditions. Epilepsy, Alzheimer's, Parkinson's, MS, autism, and even depression are all conditions in which oxidative stress runs rampant in the brain, propagating the disease process.[3]

This is why an alternate fuel source to glucose, the biological equivalent of a fossil fuel, might be of value, one that burns "cleaner" and more efficiently and can be sustained for a longer period of time. As it happens, we don't have to look very far. Scientists have known since the mid-sixties of a powerful fuel source hidden in each and every one of us, discovered upon observation of an ancient practice.

Opening Up the Ketone Firehose

Nearly every major religion has their version of a fasting protocol, from the Islamic month of Ramadan to the Jewish Day of Atonement, Yom Kippur. In the Book of Acts, from the Christian New Testament, it is said that believers would fast before making important decisions. What all of these ancient traditions have in common is that they recognized the psychological and physiological effects of fasting long before they understood the science behind it.

After a person has finished digesting the last calorie from a meal, the first source of backup fuel that the brain will tap is from the liver. The liver plays hundreds of incredibly important roles in the body—you can consider it a multipurpose high-tech manufacturing plant, capable of packing, shipping, storing, and disposing of an infinite array of important chemicals and fuels. In the previous chapter, you learned about the liver's job of recycling cholesterol carriers like LDL, but another important role is its ability to provide a small buffer of stored sugar called *glycogen*.

When blood glucose levels begin to drop, the liver releases glucose into the blood. The liver's storage capacity is fairly limited, holding only about 100 grams of glycogen. That means this backup source of sugar is short-lived, lasting only about twelve hours, give or take, depending on activity levels.

After the liver runs out of stored sugar, your brain becomes the man-eating Audrey II from *Little Shop of Horrors*, demanding to be fed. This is what most people feel when they experience the combination of being hungry and angry, known affectionately as being "hangry." This sensation is owed in part to a brain that has become the equivalent of a man-eating alien from outer space. Ever the obedient servant—and the Seymour in our *Shop of Horrors* analogy—the liver triggers a process called *gluconeogenesis*, which translates to "new sugar creation."

As nature's ultimate recycling plant, the liver kills two birds with one stone—when the body runs out of sugar, the liver will take worn-out, dysfunctional proteins from around the body, disassemble them into their constituent amino acids, and burn them off.[4] (Chopped liver? In reality, your liver is doing the chopping, dicing up proteins and turning them into sugar.) Brain fed, body cleaned. This ability of our bodies to "clean house" as a means of cellular rejuvenation is called *autophagy*, and is currently an exciting area of focus for longevity researchers.

When you experience regular periods of feeding and fasting, autophagy occurs on a daily basis. Today, unfortunately, we seldom allow it to, with the lever jammed permanently in feeding mode. But even though it is a desirable process, without a system of biological checks and balances, it could get out of hand quickly. Your skeletal muscle (like your biceps or quads, or, heaven forbid, your *glutes*) could become a target for gluconeogenesis, seeing as how they are essentially big protein "banks."

Breaking down muscle wouldn't exactly be desirable for a hungry hunter-gatherer. During periods of famine, this also wouldn't buy you very much time—to support the brain's metabolic requirements with protein alone would lead to death in about ten miserable days.[5]

To prevent this, a hormone called *growth hormone* becomes sharply elevated when the body is fasting. Growth hormone serves many roles, but its main function in adults is to *preserve lean mass in a fasted state*—that is, to stop the breakdown of muscle protein for glucose. After just twenty-four hours of fasting, growth hormone can shoot up as high as 2,000 percent (more on this in chapter 9), sending our bodies a signal to suspend muscle breakdown and rev up the fat-burning machinery instead.

Fat, on the other hand, is there to be burned. It's the body's firewood, containing more than 3,000 backup calories of brain fuel in just a single pound. An average-weight person walks around with tens of thousands of backup calories, while an obese person might carry hundreds of thousands! Unlike with sugar, the number of calories we can store as fat is virtually limitless.

When adipose tissue, the fat that sits underneath our skin and around our waists, gets broken down during times of starvation, fatty acids are released into the bloodstream to be converted by the liver into a fuel called *ketone bodies*, or simply ketones. Ketones are easily taken up by the cells of the brain and can supply up to 60 percent of the brain's energy requirements. In a paper published in 2004, pioneering ketone researcher Richard Veech wrote, "Ketone bodies deserve the designation of a 'super fuel,'" and you're about to learn why.

The Solution to Pollution?

Unlike glucose, ketones are considered a "clean-burning" fuel source because they create more energy per unit of oxygen in fewer metabolic steps, thereby generating fewer zombie molecules (free radicals) in their conversion to energy.[6] They also have been shown to dramatically increase the availability of natural antioxidants like glutathione, the body's most potent neutralizer of free radicals, making ketone utilization an anti-aging "two-for-one" deal.[7]

But the longevity-promoting virtues of ketones don't end there. Their presence in the brain has been shown to ignite gene pathways that boost levels of BDNF, the "growth hormone" that can facilitate healthy mood, learning, and plasticity, further protecting our neurons from the wear and tear of simply living.[8] As I mentioned in the previous chapter, they also positively affect the brain's blood supply, increasing blood flow by as much as 39 percent.[9]

BABY FAT ISN'T JUST CUTE—IT'S A BATTERY

Have you seen a baby lately? I'm talking about a newborn, fresh out of the womb. They're fat. And cute. But mostly fat. Packed with stored energy prior to birth in the third trimester, the fatness of human babies is unprecedented in the mammal world. While the newborns of most mammal species average 2 to 3 percent of birth weight as body fat, humans are born with a body fat percentage of nearly 15, surpassing the fatness of even newborn seals. Why is this so? Because humans are born half-baked.

When a healthy human baby emerges from the womb, she is born physically helpless and with an underdeveloped brain. Unlike most animals at birth, a newborn human is not equipped with a full catalogue of instincts preinstalled. It is estimated that if a human were to be born at a similar stage of cognitive development to a newborn chimp, gestation would be at least double the length (that doesn't sound fun—am I right, ladies?). By being born "prematurely," human brains complete their development not in the womb, but in the real world, with open eyes and open ears—this is probably why we're so social and smart! And it is during this period of rapid brain growth, what some refer to as the "fourth trimester," that our fat serves as an important ketone reservoir for the brain, which can account for nearly 90 percent of the newborn's metabolism.[10] Now you know: baby fat isn't just there for pinching. It's there for the brain.

In the context of a "normal" Western diet that's rich in carbohydrates, significant ketone production is inhibited the vast majority of the time.[11] This is because foods that are high in carbohydrates elicit an insulin response from the pancreas, and ketosis is brought to a grinding halt whenever insulin is elevated. Suppression of insulin, on the other hand, either by fasting or by eating a diet very low in carbohydrates, *drives* ketogenesis. Let's explore these two routes to ketone creation.

Intermittent Fasting

Today, humans spend most of their time feeding, with little time spent in a fasted state. We typically eat from the moment we wake up to the moment we go to sleep. This, however, was not the case for most of human history. Long before religion or diet books made calorie deprivation a carefully considered choice, our preagricultural ancestors regularly experienced fasting as a consequence of an unpredictable food supply. Their brains (and those we inherited) were forged in this uncertainty and are elegantly adapted to oscillate between the fed and fasted states as a result.

By periodically restricting our food intake, we force physiological adaptation and the production of ketones. There are many different fasting protocols that one may choose to employ. By ensuring a sixteen-hour window has passed since your last ingested calorie, you are practicing the common "16:8" method of fasting (which entails sixteen hours of fasting and eight hours of allowed feeding). This can be done daily and confers many of the benefits of fasting, namely reducing insulin and promoting the breakdown of stored fats. (We usually recommend women start with twelve to fourteen hours rather than sixteen. Women's hormone systems may be more sensitive to signals of food scarcity. For example, extended fasts may negatively affect fertility.)

One way to achieve a fast of twelve to sixteen hours may be simply skipping breakfast, a nonessential meal despite what cereal companies will tell you. Extending the fast that you endure every night while sleeping also makes use of the body's waking hormone, cortisol, which peaks thirty to forty-five minutes after waking. This hormone helps to mobilize fatty acids, glucose, and protein from storage for use as fuel, which may provide a small added bonus (more on this in chapter 9) over skipping dinner.

Skipping breakfast also works because it's often easier to start eating later than it is to stop eating earlier, since dinner tends to be our most social meal. But if you're unable to skip breakfast, eating dinner earlier is a worthy alternative, as shown in a recent study from Louisiana State University. In this trial, overweight subjects consumed all their calories between 8 a.m. and 8 p.m.—the average feeding time for most folks. But when the researchers told the subjects to skip dinner and stop eating at 2 p.m., the burning of fat (i.e., ketones), as opposed to glucose, was increased. Subjects also showed improved metabolic flexibility, which is the body's ability to switch between burning carbs and fats. This means that having a light dinner, eating it earlier in the evening, or skipping it entirely once or twice a week may help stoke the fat-burning flames. (Eating late also can disrupt the body's natural inclination to wind down at night.)

Other fasting protocols being studied include alternate-day fasting (which, like the 16:8 method, is another example of "time-restricted feeding") and periodic very-low-calorie diets. The idea behind the latter is that the body reacts to an energy deficit by releasing stored calories regardless of whether carbohydrates are consumed. This so-called fasting-mimicking diet (a diet coined by researcher Valter Longo) may confer significant benefits, including decreased risk factors and biomarkers for aging, diabetes, cancer, and neurodegenerative and cardiovascular disease.[12]

From the available options, which style of fasting should you choose? Henry David Thoreau famously observed that "life is frittered away by detail." When it comes to picking (and sticking to) a protocol, most people, male or female, will benefit from not eating for an hour or two (or more) after waking, and not eating for two to three hours before bed. This will make use of the body's natural rhythms to optimize ketone creation, among other positive things.

CREATINE: A MUSCLE (AND BRAIN) BUILDER

In the sea of overhyped marketing claims that support a billion-dollar-per-year supplement industry, creatine holds court as one of the few markedly effective tools with both a strong evidence record and safety profile. It's a natural substance produced in the body and found in red meat and fish (one pound of raw beef contains 2.5 grams of creatine), and supplementing with it leads to substantially increased muscle performance.

Adenosine triphosphate (ATP) is the energy currency of cells, used during muscle contraction. Once ATP gets used by a cell during intense exercise, creatine acts like an energy reserve, recycling it to create new ATP. No additional glucose or oxygen is needed, and ATP output remains constant. Consuming additional creatine leads to increased cellular energy stores in muscle, which allows for increased energy replenishment.

But creatine isn't just for powering through heavy gym workouts. It's a necessity in the brain, acting as a high-energy buffer to help rapidly recycle ATP. While ATP use holds stable during mental exertion, creatine levels drop in support of the energetic needs of the brain, and higher levels of brain creatine are correlated with better memory performance.[13]

Because they don't eat red meat or fish, vegetarians and vegans lack the main sources of dietary creatine, and as a result they have lower levels of it in their blood than omnivores.[14] (Though the body does create its own creatine, doing so is a stress on the system—one that can raise levels of an amino acid called *homocysteine*, which is a risk marker for heart disease and Alzheimer's.)[15] When vegetarians were given supplemental creatine (20 grams per day for five days), their cognitive function improved.[16] This was replicated in another study, where supplementation of only 5 grams of creatine per day for six weeks enhanced working memory and processing speed and reduced mental fatigue in vegetarians. According to the researchers, these findings underlined a "dynamic and significant role of brain energy capacity in influencing brain performance."

In these studies, young, healthy omnivores didn't experience a significant cognitive boost, but vegetarians did. Why? The brain may have a saturation point beyond which supplementing with additional creatine is futile, and by simply eating meat a person reaches this point. On the other hand, for those who do not consume much red meat or fish, there may be room to "top off" the brain's supply to the benefit of one's cognitive function. But low meat eaters aren't the only group who may benefit: the body's ability to produce creatine and supply it to the brain may diminish with age.[17] One surprising study in elderly omnivores found that creatine supplementation did boost cognition.[18] And, finally, carriers of the Alzheimer's risk gene, the *ApoE4* allele, have lower levels of brain creatine.[19] They, and those who are at risk of or are already experiencing cognitive symptoms, may benefit from the neuroprotective and energy-sustaining aspects of creatine. (Be sure to double-check with your physician before taking creatine supplements, especially if you have kidney problems.)

The Ketogenic Diet

The classical ketogenic diet is the gold-standard means of dramatically increasing ketone production without having to engage in time-restricted feeding or dropping calories. The diet focuses on minimizing insulin secretion with an extremely restricted carbohydrate intake, deriving 60 to 80 percent of calories from fat, 15 to 35 percent from protein, and 5 percent from carbohydrate.[20] For someone on a ketogenic diet, concentrated sources of carbohydrates, whether from sweet fruit, grains, or starchy vegetables like potatoes, are prohibited.

PROTEIN ON A KETOGENIC DIET

Contrary to popular belief, ketogenic diets are not meant to be high in protein. This is because excess protein (beyond what is required to maintain your muscle) can be transformed into glucose in the body, a process called *gluconeogenesis*. Dietary protein is also insulin-stimulating, albeit to a much smaller degree than carbohydrates—insulin helps to shuttle amino acids from the protein into skeletal muscle tissue to aid in repair (this is helpful in the context of resistance training, for example, where it promotes muscle-protein synthesis).

The ketogenic diet has been in clinical use for over eighty years as a powerful treatment for epilepsy, where it can dramatically reduce seizure incidence and calm inflammation in the brain. It's been so effective, and its safety record so robust, that it's currently being evaluated as a therapeutic option for numerous other neurological diseases. Migraines, depression, Alzheimer's, Parkinson's, and even amyotrophic lateral sclerosis, or ALS, are all conditions that have been associated with excessive brain inflammation.[21] Any of these conditions may theoretically benefit from ketones not only for treatment but in prevention. (The ketogenic diet has been found to improve memory function in patients with mild cognitive impairment—considered pre-dementia—and even early Alzheimer's disease.)[22]

Ketogenic diets are also being studied as a potential treatment for certain cancers. The cells of these cancers thrive in high-insulin environments and don't have the "hybrid technology" of the rest of the body, which means they are unable to survive off of ketone bodies. Whether this plays out long-term remains to be seen, as cancer cells are notorious for circumventing, mutating, and adapting to even the most toxic environments. At the end of the day, however, insulin and closely related "insulin-like peptides" called IGF1 and IGF2 are powerful growth factors for any cells, healthy *or* cancerous, that contain receptors for them.[23]

Whether used to treat neurological problems, to serve as a metabolic reset for those with type 2 diabetes (a ketogenic diet will, on average, halve the amount of circulating insulin and improve glucose control after just one day), or for those looking to shed a lot of fat fast, ketogenic diets hold a *lot* of promise.[24]

The Genius Plan

The Genius Plan (fully outlined in chapter 11) is no doubt a variant of the ketogenic diet. It combines intermittent fasting with low-carbohydrate eating to increase ketone availability to the brain. However, it differs from the ketogenic diet described in neurological literature in a few key ways.

For one, standard ketogenic diets are not designed to consider the burgeoning science of the microbiome, which we cover in the following chapter. The microbiome rewards us when we consume an ample and diverse array of fibrous vegetables—vegetables that contain carbohydrates, even if just a few—and thus the Genius Plan includes these foods. (These vegetables also contain important vitamins and minerals that we don't want to skimp on.)

Another key difference is in the types of fat: the sheer quantity of fat that must be consumed in the textbook versions of the ketogenic diet may make it substantially more difficult to make important brain-building considerations, like ensuring proper omega-3–to–

omega-6 ratio. The medical ketogenic diet makes no such stipulations and classically relies on foods like heavy cream and cheese to make up the bulk of calories (while the Genius Plan takes the omega-3–and–omega-6 ratio into account and adjusts accordingly).

Perhaps most important, exercise is a major aspect of any optimal brain protocol, and the one you're holding in your hands is no different. Those in "chronic ketosis" from long-term ketogenic diets can find their workout performance begins to suffer, particularly when looking to put on muscle or gain strength with high-intensity exercise. Preserving muscle is essential as we age, and is in fact directly correlated with higher brainpower.[25] The occasional post-workout high-carbohydrate meal, while not typically used for a ketogenic diet, is allowed on the Genius Plan (only once metabolic flexibility is regained) to ensure that training capacity, metabolism, hormones, and lipids stay in their optimal range. I'll provide specific details on how to approach these meals on page 319.

POST-WORKOUT CARBS: A PERFORMANCE-ENHANCING DRUG?

Carbs are not "bad"—they are just woefully misused today. If you choose to consume them, it is best to time them so that the anabolic stimulus serves a functional purpose in the body—to *enhance* your performance, not weigh it down. A best-case scenario? Refilling muscle tissue with stored sugar after a bout of vigorous exercise.

Resistance training is one of the best known means of improving overall insulin sensitivity, but the post-workout period in general has the additional benefit of turning muscles into a sponge for sugar in the blood. This is owed to the GLUT4 receptor. A channel for glucose, these receptors hide under the surface of muscle cell membranes until they start contracting, at which time they pop up to the surface. (Remember the neurotransmitter receptors from chapter 2 and how they bob up to the surface? This is exactly the same economical mechanism, elegantly reappropriated for muscles. Your DNA and genome are like an erector set—with modular and swappable parts that can have completely different functions using the same building blocks!)

Once present on the cell surface, a GLUT4 receptor will turn into a spigot that allows sugar to flow into the cell like water through an open dam. This means that for the same given amount of carbs, less insulin is needed to safely partition and dispose of it if consumed post-workout. What does this mean for you? Carbs are less likely to promote fat storage, and you will sooner bounce back into a fat-burning state. Essentially, the safest time to consume simple or concentrated complex carbs is after you've been working out. Earn your carbs!

Returning to Our "Factory Settings"

It can be daunting to begin a lower-carbohydrate diet, to say nothing of intermittent fasting—believe me, I know. When I was a kid, my mom would try every year (in vain) to get me to fast for one day in observance of the Jewish holiday Yom Kippur, which I thought was pointless masochism. I would rather have taken a trip to my orthodontist, Dr. Moskowitz, to get my braces tightened than ever skip a meal. Today, however, I can easily fast for hours.

DROPPING CARBS? EAT SOME SALT

One very frequently overlooked factor that sometimes makes people feel crummy when beginning a low-carb diet is that lowering insulin (a good thing) can deplete the body of sodium. Among its myriad of jobs, sodium helps shuttle vitamin C into the brain, where it is used to create neurotransmitters that can affect your mood and memory. Sodium is also key to maintaining exercise performance as you drop the carbs.

According to cardiovascular research scientist and sodium expert James DiNicolantonio, during the first week of carbohydrate restriction, you may require up to an additional 2 grams of sodium—about a teaspoon of salt—per day to feel optimal, which can be reduced to 1 gram after the first week. Remember: individual experimentation is key. (You can watch a thirty-minute-long interview I did with James at http://maxl.ug/james dinicinterview to learn more about this fascinating topic.)

"But my doctor told me to be on a low-sodium diet for blood pressure!" Insulin and sugar may affect blood pressure more than salt. They stimulate the body's fight-or-flight cascade, which can promote high blood pressure and cause the body to hold on to more sodium anyway.

When we chronically deny our brains any respite from glucose, this creates addiction, which explains why the sudden removal of carbohydrates can lead to headaches and fatigue. This is what I experienced as a pizza- and Pop-Tart-eating preteen. However, when you combine periodic low-carb phases with regular fasting, you set the physiological stage to return your metabolism to its "factory setting." By reducing insulin and allowing the ketone firehose to be turned on, you regain metabolic flexibility, thereby training your metabolism to work for you, not the other way around. This is the holy grail of metabolic health.

The following seven steps to becoming metabolically flexible all involve adapting your brain to use ketones from fat for fuel, and they mimic the cascade set forth by fasting. We theorize that the "hanger" and headaches that may occur during this three- to seven-day period correspond with the brain's upregulating of the enzymes required to process ketones as fuel.

These times are approximations assuming a non-keto-adapted state.

1. Depletion of last consumed carbohydrate. (4 to 12 hours)
2. Depletion of the body's stored carbohydrate. Recall that the liver can store about 100 grams of carbohydrate in the form of glycogen, give or take depending on body size. (12 to 18 hours)
3. Decrease in amino acid breakdown to preserve muscle. (20 to 36 hours)
4. Breakdown of amino acids for gluconeogenesis. (24 to 72 hours)
5. Increase in ketone production and utilization. (48 to 72+ hours)
6. Upregulating of ketone-burning enzymes in the brain. This takes up to a week but can be shortened by emptying carb stores faster with high-intensity exercise, eating an overall lower-carbohydrate diet, or integrating *medium-chain triglycerides*, discussed momentarily. (1 to 7 days)
7. Enter a state of metabolic flexibility. Here, the occasional carb meal can be consumed without interrupting the fat-adapted state, especially when consumed during or after a workout.

The key to experiencing true freedom from food lies in cutting off our dependency on glucose and reestablishing the kind of metabolic flexibility our ancestors knew well. After a few days on a low-carb diet, the feeling of hunger and craving for more carb-rich foods will begin to subside and ultimately fade away. Here are some signs that your body's fat pipeline is up and running:

- You can go several hours without eating and not want to kill anybody.
- You do not crave starchy or sugary snacks in between meals.
- Your mind is sharp and clear and your mood and energy levels are stable.
- Moderate exercise doesn't induce ravenous hunger or fatigue.

DOCTOR'S NOTE: FEMALES AND SUPER-LOW-CARB DIETS

While we generally advocate a lower-carb diet than the standard American fare, it should be noted that there are substantial genetic and gender differences in carbohydrate tolerance. In particular, females on ultra-low-carb, ketogenic diets can experience stalled weight loss, mood issues, and disruption of their menstrual cycles. The optimal amount of carbohydrate on a daily or weekly basis should generally vary based on level of physical activity, and can range anywhere from 30 to 150 grams per day. We will talk more about carb amounts and timing in chapter 11.

Ketones: A Life Raft for the Aging Brain?

Now that you know how to enter a state of ketosis, you should be aware that the benefits of allowing the brain to "burn" ketones instead of glucose don't end at the fact that they are a cleaner-burning fuel source. One of the major benefits of supplying ketones to the brain that I have yet to mention is that certain brains may actually work *better* when given the chance to run on ketones. These brains may be unable to process glucose effectively, but are given little to no alternative thanks to our "keto-deficient" diets, a term coined by ketone researcher Sam Henderson.[26]

A prime example may be carriers of the most well-defined Alzheimer's risk gene, the *ApoE4* allele. Carriers of either one or two copies of this gene, which make up more than a quarter of the population, have been demonstrated to exhibit low glucose metabolism in the brain.[27] This occurs seemingly across the age spectrum, starting as early as the twenties and thirties, which is *well* before the age at which memory-related symptoms typically emerge.

Carriers of the *ApoE4* allele have either a two- or twelvefold increased risk of developing the disease depending on whether they've inherited one or two copies. Despite the increased risk, however, many *ApoE4* carriers never go on to develop Alzheimer's disease. Weirder still, a significant number of patients with Alzheimer's disease do not carry this gene variant at all. And yet, noncarriers who develop Alzheimer's disease ultimately display the same reduced glucose metabolism in their brains as *ApoE4* carriers, implicating impaired brain-glucose metabolism as a possible causative factor in the disease. This paradox begs the question: is the worrying relationship between *ApoE4* and Alzheimer's disease yet another symptom of the eating pattern we've been coerced to adopt?

The *ApoE4* gene is considered the "ancestral" gene, having been in the human gene pool longer than other variants. In populations with earlier exposure to agriculture (i.e., access to grains and starches), the frequency of the gene is lower, suggesting that our modern diets may have selected against carriers of this gene.[28] Even today, when we look to less industrialized parts of the world, the theory holds water. Take the Yoruba people of Ibadan, Nigeria, whose diets have not been industrialized the way ours have. Among them, the *ApoE4* gene is relatively common, and yet has little to no association with Alzheimer's disease when compared to African Americans.[29] The Yoruba tend to consume less than one-third the sugar we Americans do per capita, and lower-glycemic-index carbohydrates in general.[30]

What does this mean for you? If you carry the Alzheimer's-risk gene (statistically, one in four readers of this book will) your brain may be particularly ill suited to the "post-agricultural" high-sugar, high-carbohydrate diet.

By the time a person is diagnosed with Alzheimer's disease, brain glucose metabolism is already reduced by 45 percent compared to healthy people. But as I've mentioned, any brain can experience difficulties deriving energy from glucose, well before memory problems emerge. Aside from the *ApoE4* allele, this situation may be the result of the same diet and lifestyle stresses that encourage the development of type 2 diabetes.[31] In one eye-opening study, insulin resistance in the body predicted diminished glucose metabolism (called *hypometabolism*) in the brains of cognitively normal adults. "Hallmarks of Alzheimer's disease, such as glucose hypometabolism and a loss in brain tissue, are nevertheless strongly associated with peripheral insulin resistance," wrote researchers in the journal *Physiological Reviews.* (Revisit page 103 to learn how to assess your own sensitivity to insulin.)

THE NEW FRONTIER: TREATING ALZHEIMER'S AS A METABOLIC DISEASE

When it comes to the most common form of dementia, there are likely many interacting variables that decide one's fate. As my friend Richard Isaacson, Alzheimer's prevention specialist, says, "Once you've seen one case of Alzheimer's disease, you've seen one case of Alzheimer's disease." The complexity of the disease, coupled with the fact that it begins in the brain well before the appearance of symptoms, may explain why Alzheimer's drug trials have a 99.6 percent failure rate. And why no one has ever recovered from it.

Recently, the Buck Institute for Research on Aging reported that it was able to "reverse" symptoms in nine of ten patients with varying degrees of cognitive impairment, including Alzheimer's disease. The program was designed to improve metabolic health: blood sugar and insulin levels were reduced, and patients were told to eat "low-grain" diets to spur ketone production.[32] At the same time, other factors known to play a role in metabolic health, such as nutrient deficiencies, sleep problems, and sedentary lifestyles, were addressed. In total, thirty-six customized interventions were "prescribed" to each subject, many of which are in line with the recommendations made in this book.

At the end of six months, most patients reported improvements in their ability to think and remember, which their partners corroborated. Cognitive testing revealed an improvement as well. The report stated that some who had been unable to work due to the severity of their cognitive decline were able to resume their jobs, and brain scans even showed that one patient added new volume to the vulnerable hippocampus—a growth of nearly 10 percent!

Does this mean that *Alzheimer's disease* might be "reversible"? While it's tempting to draw big conclusions from these few anecdotes, only a handful of patients in this trial had actual Alzheimer's disease. Therefore, to answer this question, larger controlled trials with more rigorous scientific methodology would be needed. Nonetheless, this "kitchen sink" approach has presented a new and worthy angle from which to tackle cognitive impairment: as a metabolic problem.

Taken together, it's perhaps not surprising, then, that a research group out of Brown University (led by neuropathologist Suzanne de la Monte) coined the term "type 3 diabetes" to describe Alzheimer's disease. This concept, which has since been referenced widely throughout the medical literature, directly characterizes Alzheimer's disease as being metabolic in origin.

Make no mistake about it: an energy-deprived brain is bad news.

In fact, the forgetfulness that we associate with typical aging may be among the first signs that the brain is struggling to fuel itself. The good news is that aside from helping to reduce oxidative stress and inflammation, providing ketones to the brain (with the Genius Plan or any variant of a ketogenic diet) may help the brain "keep the lights on" well into old age. This is because unlike with glucose, a brain's ability to derive energy from ketones is seemingly not affected by old age, the *ApoE4* gene, or even Alzheimer's disease.[33]

As a bonus, ketogenic diets have even been shown to increase the number of brain mitochondria (the power plants of cells)—thereby increasing metabolic efficiency, which otherwise diminishes with age and more markedly in neurological conditions.[34]

Can't I Just Eat My Ketones?

There is another means of supplying ketones to the brain that I have briefly mentioned: by consuming special ketone-generating foods. These are foods that contain a natural source of a relatively rare dietary fat called a *medium-chain triglyceride*, or MCT. MCTs are abundant in coconut oil, palm oil, goat's milk, and human breast milk, and they have a unique and important effect on the body. Upon consumption, these fats go straight to the liver* to become converted to ketones, an astonishing property that can elevate the amount of ketones in the blood, day or night, fasted for fed.[35] Researcher Stephen Cunnane found that in a non-ketotic, non-fasted state, the brain could potentially derive 5 to 10 percent of its fuel from these supplementary ketones. Interestingly, this is the same degree of fuel lost to the hypometabolism seen in young brains with the *ApoE4* allele.

Of the 14 grams of fat in a typical tablespoon of coconut oil, 62

---

* Most fats, like the fat from olive oil or a grass-fed-beef hamburger, enter the lymphatic system upon consumption, where they get disseminated throughout the body.

to 70 percent is pure MCT, most of which is lauric acid. In breast milk, lauric acid also makes up the largest proportion of MCTs. On top of lauric acid, coconut oil contains other fatty acids including capric acid and caprylic acid, which may be even more ketogenic—particularly the latter, which is the main fatty acid advocated for use in the treatment of drug-resistant epilepsy.[36] Often, these fatty acids are isolated to create formulations of pure MCT oil, which are nearly 100 percent ketone-producing medium-chain triglycerides.

FAQ: Do ketone supplements and MCT oil help me burn more fat?

A: MCT oil and ketone supplements may provide serious cognitive benefits, but they are often marketed as a weight-loss aid—for which they are not ideal. Ketones made by your body are the by-product of fat being burned. When you add exogenous ketones, which is still a form of energy that will need to be burned, you're actually **preventing** your body from using its own fat. For weight loss, we think it's better and more meaningful to make your own ketones than to consume them from outside sources.

For people with Alzheimer's disease or other neurodegenerative conditions, MCT oil may be particularly useful. In Alzheimer's disease, a shift in food preference occurs where sufferers develop a sweet tooth.[37] This may be a distress call from their metabolically ailing brains, starving for energy, to attain sugar in the form of rapidly digestible carbohydrates—the exact kinds of foods that spike insulin, drive inflammation, and block ketone production. Supplementary coconut or MCT oil theoretically may circumvent this particular issue while dietary carbohydrates are gradually lowered. When these dietary ketones are given to sufferers of memory loss, cognition has improved in some studies, and at least one case report has detailed positive responses from a patient with advanced Alzheimer's who consumed as little as two tablespoons per day (see sidebar).[38] You can even get a prescription for a medical food made from caprylic acid, which is FDA-approved for the treatment of Alzheimer's disease. (Its use as a preventative strategy for Alzheimer's is currently being studied.)

MARY NEWPORT: A COCONUT OIL PIONEER

I became familiar with Mary Newport's work with coconut oil early on. Her husband, Steve, was diagnosed with Alzheimer's disease and became unable to perform many of his daily activities, including some of his favorite pastimes. After trying all available pharmaceutical options without much luck, Mary set out to look for something better.

She stumbled upon a press release for a "medical food" being developed that was composed of caprylic acid, an MCT. The press release claimed that by supplying ketones to the brain, it improved memory and cognition in nearly half the Alzheimer's patients tested. With the disease now seven years advanced and quickly ravaging her husband's brain, Mary was desperate to get her hands on this experimental product, but FDA approval would not happen for another year. That's when she had a revelation.

Mary Newport happens to be a physician in neonatology, a subspecialty of pediatrics that consists of the medical care of newborn infants. She was already familiar with MCTs because they are a component of breast milk and were commonly used in the seventies and eighties to help very premature newborns gain weight. Since then, MCT *and* coconut oil have been added to virtually all infant formulas. Mary's unique training taught her that she might simply be able to give her husband coconut oil instead of waiting for the medical formulation to make it to market.

Mary began giving Steve just over two tablespoons of coconut oil a day, calculated to the equivalent amount of MCTs in a dose of the medical food, and then administered a clock-drawing test, which is sometimes used to assess cognitive function (anyone with a loved one with dementia will be all too familiar with this test). After just two weeks on daily doses of coconut oil, Steve's clock drawing improved dramatically. Soon, Mary began cooking with it and giving it to Steve any chance she could. By the fifth week, Steve's clock was night-and-day better than on the first day.

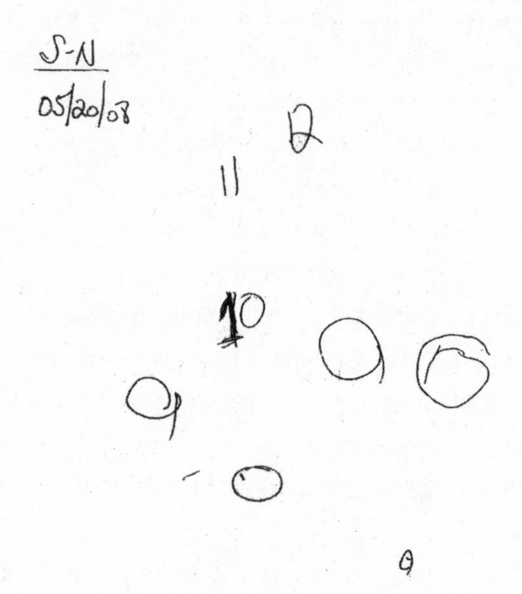

1 DAY BEFORE COCONUT OIL

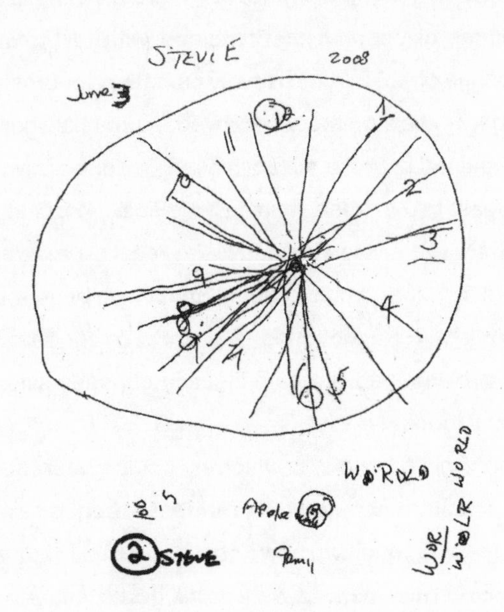

14 DAYS ON COCONUT OIL

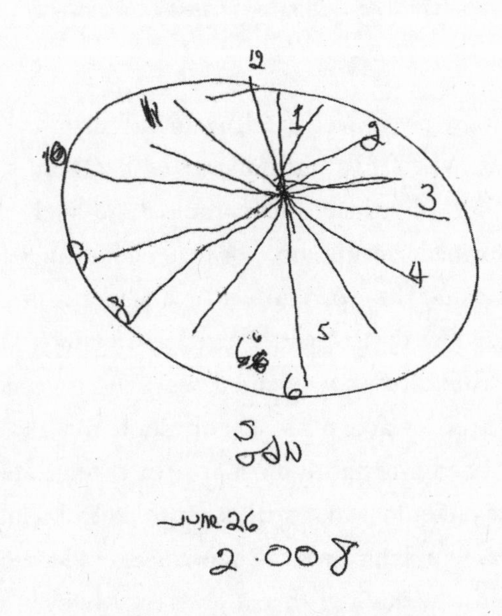

37 DAYS ON COCONUT OIL

Over the following year, Mary gradually increased Steve's dose up to eleven tablespoons of coconut oil combined with MCT oil per day (increasing the dosage of MCT oil too rapidly can cause diarrhea). According to Mary's reports, Steve's memory improved, as did his scores on cognitive tests. He regained many of his abilities to perform daily tasks, "going back in time at least two to three years in his disease process," says Mary. On two days that Steve missed his dose, she recalls a marked regression in his abilities—a sign that the coconut oil may have been responsible for Steve's improvement. She continued to give Steve coconut oil for nearly a decade, and implemented diet and lifestyle changes similar to those I recommend in this book.

Mary's coconut oil experiment, which has since been published as a case study in the journal *Alzheimer's & Dementia*, began seven years into her husband's disease, which we now know had actually started in Steve's brain decades prior. Ultimately, Steve lost his battle with Alzheimer's disease and passed away in 2015, but his story continues to live on in Mary's advocacy.

For those that are cognitively healthy and accustomed to a high-carbohydrate diet, MCT oils and fats *might* also help supplement brain energy while you reduce your carbohydrate load—but this is currently just informed speculation. For many adopting a lower-carb diet for the first time, the "low-carb flu" can be a common occurrence for the first few days, characterized by feelings of lethargy, brain fog, and irritability—so anything with the potential to help give your brain a boost and push you through this period would be worthwhile. There's certainly no harm in trying, and some of the brain benefits of ketones may still be attained by including these oils in your diet as you reduce your dependence on glucose. Feel free to experiment—just be advised that it hasn't really been tested yet. (And don't forget to add salt, which I mentioned on page 158.)

While you may be thinking that MCT oil would be the perfect topping for a serving of pasta or added to your morning bowl of cereal (who wouldn't want to have their cake and eat it too?), forcing an elevation of ketones in the context of a high-carbohydrate diet ignores many of the underlying problems driving neurodegeneration and brain aging to begin with—that is, excessive insulin. Equally important, the transient ketones from a supplement will never reach the concentrations achieved during a ketogenic diet or period of fasting. Supplementing with coconut oil or even with commercially available pure MCT oil in the non-fasted state is the equivalent of pouring water into an already-full cup. Fasting or eating a low-carb diet, thereby allowing your body to generate its own ketones, is akin to drinking from that cup instead.

Remember: "biochemical liposuction"—when your body is able to tap into its own fat stores and use those fats for energy—occurs when you drop your insulin sufficiently, not because you've added more fat into your diet. In healthy adults, ketones are simply a marker of all the other wonderful processes that go along with fasting that we've already mentioned. When you're burning your own fat, adding oil contributes calories—which is fine, but keep in mind that a dramatic surplus of calories, whether from carbs, protein, or fat, can eventually cause you to gain weight. In modern times, many of us have gone most of our lives without our brains adapting to using our own fat as fuel, because we are always eating. Give your body and brain a chance to burn that fat and your biology will repay you in spades.

Up next, a "forgotten organ" hidden within you—and its powerful role in the healthy performance of your brain.

FIELD NOTES

- Ketones are considered a "super fuel," capable of reducing oxidative stress in the brain and upregulating genes involved in neuroplasticity.
- Certain brains are unable to use glucose effectively, and ketones may provide an alternate fuel source.
- A common misconception is that ketones are created as a result of eating more fat. In actuality, ketones are generated when insulin is reduced, which is the result of fasting or a low-carbohydrate diet.
- Metabolic flexibility is a greater goal than chronic ketosis (unless treating a neurological condition whereby medical ketosis may be warranted). With metabolic flexibility we can enjoy dalliances with ketosis while also nurturing gut health and enjoying the occasional carbohydrate "fuel-up" to sustain physical performance. This won't interrupt the fat-adapted state.
- Raising ketones with MCT oil while consuming excessive carbohydrates defeats the purpose and ignores many of the underlying problems driving neurodegeneration.

GENIUS FOOD #6

GRASS-FED BEEF

The meat industry as it currently stands is cruel, unsustainable, and frankly indefensible. In the case of beef, the industry produces meat that is unhealthy, from stressed-out animals that are pumped full of antibiotics and fed a highly unnatural diet of throwaway grains and even candy.* But let's not conflate factory-farmed beef with the beef that comes from healthy cows that have been allowed to pasture on grass (their natural diet), experiencing—as their farmers like to say—only one bad day.

Much of the debate surrounding the nutritional value of meat centers on protein, but I believe it's critical to broaden the conversation to nutrients *other* than protein that play an important role in our cognitive function. For example, grass-fed beef is a rich source of essential minerals like iron and zinc, where they are packaged in a form that the body can easily utilize. (This is unlike, say, the iron from spinach or zinc from legumes.)[1] Grass-fed beef is also a great source of omega-3 fats, vitamin B_{12}, vitamin E, and even certain nutrients, such as creatine (covered on page 153), which, though not essential, are highly beneficial. Researchers believe that it was access to these very nutrients (along with the burst of caloric energy from cooked meat) that catalyzed the evolution of our brains into modern cognitive supermachines. Deficiencies in any of these micronutrients are linked

* You read that correctly. Feedlot animals are routinely fed junk food like candy, cookies, and marshmallows because those foods provide cheap carbs to fatten them up.

with brain-related disorders, including low IQ, autism, depression, and dementia.

Few know the link between diet and mental health better than Dr. Felice Jacka, director of Deakin University's Food and Mood Centre, and someone whom I've had the privilege of interviewing. In 2017, she published the world's first randomized control trial showing the antidepressant effect of healthy food (I detail her findings on page 224). Previously, she found that women who didn't eat Australia's national recommendation of three to four servings of beef per week were *twice as likely* to be depressed, or suffer from an anxiety or bipolar disorder, as those who did.[2] (She also found that while some was better than none, more wasn't necessarily better—women who consumed more than the recommended amount were also at increased risk.) In Australia, cows tend to be grass-fed by default—an important caveat.

What about the value of meat to the cognitive function of a particularly vulnerable group: children? Far from the reach of food delivery apps, malnourishment still poses a public health problem in various parts of the world. One of these places is Kenya, where Charlotte Neumann, a researcher at UCLA's Fielding School of Public Health, observed that children who consume more meat actually tend to perform better, physically, cognitively, and behaviorally. To see what effect, if any, meat consumption might have on the developing brain, Dr. Neumann designed a trial.

She divided children from twelve Kenyan schools into four groups. One group served as the control, while children in the other three groups each received a porridge made up of corn, beans, and greens every morning for breakfast. One group received the mixture with a glass of milk, another had ground beef added, and the third got just the plain-Jane version. All versions were balanced to contain the same number of calories, and the study ran for two years.

Compared with the other groups, students in the meat group gained more muscle mass and had fewer health problems than the children who consumed the porridge plain or with milk.[3] They also showed greater confidence in the playground—a sign of improved mental health. Cognitive performance was stronger too. While all groups improved, the meat group showed the steepest rate of improvement in math and language subjects. Neumann and her colleagues wrote:

> The improved cognitive performance and increased physical activity and leadership and initiative behaviors in the meat group may be linked to greater intake of vitamin B_{12} and more available iron and zinc as a result of the presence of meat, which increases iron and zinc absorption from fiber- and phytate-rich plant staples. Meat, through its intrinsic micronutrient content and other constituents and high-quality protein, may facilitate specific mechanisms, such as speed of information processing, that are involved in learning.

This study was performed in children, but we now know that our brains continue to change throughout life—supplying them with the nutrients they need should be top priority. Still, many will write off all meat as unhealthy, but to this I say (quoting Carl Sagan): "Extraordinary claims require extraordinary evidence." Meat and the nutrients it contains were an essential part of the evolution of our brains, with evidence of butchery by early humans dating back more than three million years.[4] Today, we have the luxury of getting to choose our meals based on ethics, but our forebears had no such privilege; they would not have passed up the opportunity for the life-sustaining nutrients contained in fresh meat. The notion that properly raised animals, providing a bevy of highly bioavailable nutrients, are somehow bad for us would be an extraordinary claim, with little good evidence to back it up.

While I'll never know if my mom's lifelong abstinence from red meat had anything to do with her memory loss or the bouts of depression she would occasionally suffer during my childhood, it's clear that it hadn't protected her either.

How to buy: Look for humanely raised 100 percent grass-fed and grass-finished beef, ideally organic and from local farms. Note that organic beef, unless it clearly states "100 percent grass-fed," is usually from cows fed organic grain.

Pro tip: Ground grass-fed beef tends to be much more economical than chops. If access to grass-fed beef is difficult, mail-order subscription services are a great alternative. You can find some recommendations at http://maxl.ug/GFresources.

How to cook: While grass-fed beef has triple the vitamin E of grain-fed beef, which helps protect its polyunsaturated fats from oxidation, I recommend using as low a heat as possible. Consider cooking with garlic- and onion-based marinades to reduce the formation of neurotoxic compounds like heterocyclic amines.[5] Always pair with fibrous veggies such as kale, spinach, or Brussels sprouts, which help to neutralize oxidative products in the gut, and avoid consuming with starchy vegetables, grains, and other concentrated carbs.

Bonus points: Eat organ meats and drink bone broth! Both are full of important nutrients not contained in muscle meat, such as collagen. Collagen contains important amino acids, which too have become lost to the modern diet. One of them, glycine, has been shown to improve sleep quality and may increase brain levels of serotonin (important for healthy mood and executive function).[6]

CHAPTER 7

GO WITH YOUR GUT

If you want to go fast, go alone. If you want to go far, go together.

—**AFRICAN PROVERB**

We humans have known since time immemorial something that science is only now discovering: our gut feeling is responsible in no small measure for how we feel. We are "scared shitless" or we can be "shitting ourselves" with fear. If we don't manage to complete a job, we can't get our "ass in gear." We "swallow" our disappointment and need time to "digest" a defeat. A nasty comment leaves a "bad taste in our mouth." When we fall in love, we get "butterflies in our stomach."

—**GIULIA ENDERS, *GUT***

If you're like most people, the thought of a collection of *trillions* of bacterial cells living inside of you may be enough to make you want to run for the closest shower. The implicit "ick factor" is compounded by the fact that we live in a culture that seizes every opportunity to sell us antibacterial soaps and sanitizers. But the truth is, we've been sold a lie about bacteria: without them, we wouldn't be here.

You're now familiar with mitochondria, the cellular organelles responsible for combining glucose (or ketones, a by-product of fat metabolism) with oxygen to create energy. These important structures didn't always work for us. As the theory goes, they were once bacteria floating about the world, when one mitochondrion became engulfed by another bacterium. Rather than digest it, the much larger host cell was able to exploit its new friend's energy-producing capabilities for survival—a serious advantage 1.5 billion years ago as the world was becoming increasingly oxygenated. In return, the mitochondrion got protection from the elements and an unlimited all-you-can-eat buffet—but they could never, ever leave. It was perhaps the earliest-ever case of Stockholm syndrome.

Over time the mitochondrion and its host cell began to depend on each other, joining the ranks of famous partnerships like Batman and Robin, Han Solo and Chewbacca, and Bert and Ernie (well, maybe not *quite* like Bert and Ernie). This was the birth of the complex *eukaryotic* cells that ultimately gave rise to multicellular organisms like us. Even after all these years, it's striking to realize that our mitochondria can still multiply inside our cells and hold on to their own, completely separate set of DNA—a throwback to their lives as bachelorettes.

We would be nowhere without bacteria. And though our modern form is vastly more complex than in those early years as single-celled organisms, our communion with bacteria today is no less important. There are countless microbes on our skin, around our ears, in our hair, in our mouths, on our genitals, and in our guts. Even parts of ourselves once thought to be sterile, like the lungs and the mammary glands of the breasts, are now known to be posh country clubs for microbes.[1] Each location has its own population that lives in symbiosis with its unique environment. The gut microbe population, for example, which consists primarily of bacteria that live without oxygen, would die instantly if placed next to the microbes on your face, which delight in their exposure to fresh air.

The general term that we use for the cumulative genetic content of all of these simple, single-celled organisms is the *microbiome*. Your house has its own microbiome, which represents the genetic material carried by the microbes that live in it. Your home microbiome may differ dramatically from your neighbor's depending on whether you have a dog or a small child, and whether you live in a city or in a suburb. Even entire cities have their own microbial signatures.[2] The microbiome of Los Angeles, for example, is different from that of New York. Do West Coast bacteria prefer on-camera work to the stage, where the East Coast microbes *really* shine? These questions remain to be answered by science.

Though you have microbes all over the exterior of your body, the vast majority of microbial cells that you carry with you reside in your gut. This is your gut microbiome. Though we once believed they outnumbered our own human cells ten to one, we've since arrived at a more accurate estimate—about thirty trillion—placing their population roughly at the same number of cells that contain human DNA. No less impressive, however, is that the weight of these bacteria alone equals the weight of your brain—somewhere between two and three pounds!

WHAT'S IN YOUR POOP?

The average stool sample is more than half bacteria, with each gram consisting of one hundred billion microbes. That's nearly fourteen times the global human population in just one gram of dookie! Fecal matter is so dense with microbes, in fact, that every time you go to the bathroom, you excrete about one-third of your colonic bacterial content. Not to worry, however, as the colonic bacterial count rebuilds over the course of the day.[3]

These microbes contain lots of information, each carrying their own unique genetic material. If we consider the total amount of genetic material represented by our bacterial friends—whose DNA length is typically one to ten megabases, which holds one million bytes of information—just one gram of human stool has a data capacity of one hundred thousand terabytes! And you thought your thumb-drive keychain was cool.[4]

Much like we outsource aspects of our cognition to our smartphones—the ability to remember phone numbers, for example, freeing up our brainpower for other tasks—we outsource many services to our microbiome. It's able to provide these services in part because it represents genetic material nearly one hundred times as complex as our own (relatively rudimentary) genome of twenty-three thousand genes. This makes the microbiome capable of a wide range of functions—from keeping our immune systems healthy to extracting calories from food to synthesizing important chemicals like vitamins.

It might not seem obvious, but the gut and brain have a very close relationship as well. Our microbiome is connected to our moods and behavior, communicating with our brain through the vagus nerve, which provides a direct line between brain and gut, as well as through the various chemicals it produces and releases into our bloodstream. The rent paid by our bacterial immigrant population is so underappreciated, it's no wonder scientists are now referring to this writhing mass of genetic material as our "forgotten organ."

MTV Cribs: Microbiome Edition

Though we'd rather not think of ourselves as elaborate digestive tubes with legs, that's essentially what we are. Nearly every feature of our being has evolved to help us better procure energy in the form of food.

The gut, the term given to this long and windy tube also known as the alimentary canal, begins at the mouth and ends, well, you know where. Gut health and function are not usually an easy topic for people to talk about. After all, our gut makes strange noises, is a source of physical discomfort for many, and excretes things that I'm betting most would prefer to not even think about. The gut also mediates our relationship with food, which can become distorted and deranged when we are dealing with weight problems.

A journey south, beginning at the mouth and traveling down the esophagus, first reveals the stomach, followed by the small intestine, and finally the large intestine—otherwise known as the colon. Each of these segments has its own unique climate, just as you might experience when traveling from the northeast of the United States down to the sunny beaches of south Florida. As you venture farther south, you notice a difference in vegetation, foliage, birds, and insects—all naturally selected for adeptness according to temperature, local cuisine, seasonal variation, and countless other variables unique to each location.

Similarly, the gastrointestinal tract has a different climate the farther down one travels, and microbes know this. The stomach is too acidic for microbiota to reside (unless you regularly take an acid-blocking medication like millions of Americans do, which can cause many unintended and unpredictable consequences), and the small intestine, as an active site of nutrient absorption, is too volatile. There are still microbes this early in our journey—per gram of content in the stomach and small intestine, there are roughly 10^3 to 10^8 bacteria. Though they are innocuous in these quantities, problems can arise when bacteria overpopulate here. In the small intestine, SIBO, or small intestinal bacterial overgrowth, can cause bloating, abdominal pain, and even nutrient deficiencies for the host. Once we get down to the colon, however, it provides the most suitable atmosphere for these bacteria—and the concentration of microbes there shoots up to 10^{11} bacteria per gram. It's the Miami of the GI tract.

Part of the reason these bacteria are so high in number in the large intestine is that that's where your "tenants" expect to find an abundant source of sustenance. You see, the gut microbiome is composed of a kind of bacteria called *commensals*, which comes from the Latin word *commensalis*, meaning "sharing a table." They've earned this name because every time we eat, they wait silently to be fed, like thirty trillion obedient dogs. But what do *they* eat?

Thrown into a modern restaurant, commensal bacteria would

skip the menu entirely and head straight for the salad bar. It's there that these little critters would find the food that they love to chow down on: plant fibers. These fibers provide a form of carbohydrate that is inaccessible to us and passes through the stomach and small intestine undigested. When those fibers finally make it down to the large intestine, the microbes get to experience the equivalent of a Thanksgiving dinner!

MEAT AND THE MICROBIOME

A study published a couple of years ago sent shock waves down the spine of many a health-conscious meat eater. Researchers studying mice found that some species of gut bacteria consume the amino acid carnitine, found in red meat, which in turn elevates a compound called *trimethylamine-N-oxide*, or TMAO.[5] TMAO is thought to contribute to atherosclerosis, the disease process that leads to plaque-ridden arteries. The ensuing fear was that meat, independent of its saturated fat or any of the other health claims previously waged against it, might now promote heart disease via an entirely new mechanism—microbial fermentation.

Closer inspection into the research reveals a few important details. First, the mice were fed very high doses of supplemental carnitine. This caused a shift in the microbiome and gave these TMAO-producing bacteria a competitive edge in the large intestine. Second, low-grain vegan and vegetarian diets appear to select against carnitine-loving gut flora, a fact highlighted by microbiome researcher Jeff Leach.[6] In the human arm of the study, researchers were able to convince a vegan to consume an eight-ounce steak to see what it would do to her TMAO levels, and they didn't budge. Though a small "n of 1" experiment, the results suggest that to a certain point, the overall composition of the microbiome is more important than the individual food being consumed. A reasonable takeaway? Don't give meat a TMAO-out—just eat mostly vegetables and skip the grains.

Forced to adopt the modern diet, the host-microbe relationship can become strained. As I mentioned, they only really like to consume one thing—fiber. Specifically, a form of fiber called *prebiotic fiber.* This includes soluble fiber and a form of indigestible starch called *resistant starch.* Offer your standard American breakfast of refined-flour pancakes, bacon, and cheesy eggs, and your typical intestinal bacterium would politely decline.

Now, you may want to write off your microbe friends for being too picky when it comes to dining plans, but keep in mind that for hundreds of thousands of years, humans ate diets that were rich in fiber, with scientists estimating an intake of roughly 150 grams of fiber per day. Today we consume just 15 grams per day on average. Much like our inadequate consumption of omega-3s and other essential nutrients, prebiotic fibers have largely been stripped from our Western dietary pattern. The disappearance of these *microbiota-accessible carbohydrates* (a term coined by prominent Stanford University microbiologists Justin and Erica Sonnenburg) presents serious downstream health consequences, as you'll learn. But increasing the presence of these gut-happy carbs is easy, as there are plenty of foods full of prebiotic fiber. These include: berries, leeks, jicama, kale, sunchokes, avocado, spinach, arugula, garlic, onions, coffee, chicory root, unripe bananas, raw nuts, fennel, okra, bell peppers, broccoli, radishes, dark chocolate, and sprouts.

Now that you know where to find these nourishing fibers, the next few pages will close the loop, definitively linking them to improved mood, cognition, and longevity.

A Fountain of Youth

Unless you are a *Turritopsis dohrnii*, aging well is likely a concern for you. If you are one of these newly discovered "immortal" jellyfish lazing about the Mediterranean Sea, you possess the ability to regress to earlier stages of development at will. If you are not one of these lucky creatures, however, you'll likely want to preserve mind and body for as long as possible.

One result of fiber consumption is that our microbial friends metabolize the fiber and transform it into chemicals called *short-chain fatty acids*, or SCFAs.[7] This category of fatty acids includes butyrate, acetate, and propionate, all of which have been linked to many longevity- and health-promoting effects. These fatty acids are the literal waste products of bacteria, and yet we are indebted to them for it.

The most extensively studied of the SCFAs is called *butyrate*. Grass-fed beef and dairy contain small amounts of this fat, but a much more appreciable amount of butyrate is churned out by our microbiota when we simply eat more fiber. This is desirable partly because butyrate has been shown to raise levels of BDNF, which directly promotes neuroplasticity and slows neurodegenerative processes.[8]

Aside from increasing BDNF, the brain's anti-aging "Miracle-Gro," one of butyrate's most beneficial effects is in reducing inflammation. Generally speaking, the more fiber you consume, the more closely your microbiota begin to resemble an inflammation-extinguishing butyrate factory.[9] In terms of your cognitive function, less inflammation means that you can think more clearly and focus and remember things better.[10] But while taking steps to reduce inflammation is key for thinking and performing at your best, doing so may also guard you against the march of time.[11]

When it comes to longevity, the important thing to focus on is your *health* span. Unlike your life span, which describes the number of years in your life, your health span represents the amount of life in your years. Having a longer health span means less disability, better cognitive function, better mood, and being free of chronic disease for as long as possible. Ideally, we'd like long life spans and health spans to match. Unfortunately, today, our life spans are increasing (thanks in part to the wonders of modern medicine), but our health spans are not. We're simply living sick, longer.[12]

But there are some exceptions to this: people who seem to stay vibrant and healthy up until the very end of their lives. In one study that followed more than 1,600 adults for an entire decade, those who ate the most fiber were 80 percent more likely to be free of hypertension, diabetes, dementia, depression, and disability than low-fiber consumers.[13] In fact, fiber consumption determined healthy aging more than any other variable studied, including sugar intake. Not bad for a nutrient most famous for helping our grandparents more easily go to the bathroom—and that's not the end of it.

FECAL MICROBIOTA TRANSPLANTATION (FMT)

Although the transplantation of fecal material from one person to another may not be pleasant to think about, imagine for a second you had an infection called *C. difficile*. A pathogenic and antibiotic-resistant bacteria that causes profound diarrhea and intestinal inflammation, *Clostridium difficile* results in half a million hospitalizations and thirty thousand deaths annually, according to the latest estimates from the CDC. Highly opportunistic, *C. difficile* is contagious and already present in 2 to 5 percent of the human adult population. Antibiotic use is a major risk factor for *C. difficile* because these drugs decimate healthy gut populations, allowing the pathogen to exploit weaknesses in the microbial community until it becomes a full-blown infection.

In 2013, scientists wanted to see if transplanting the microbiome of a healthy person into one with such an infection could reestablish order and help defeat *C. difficile* naturally. Performed via fecal microbiota transplant (FMT), where the bacteria-rich stool of a healthy person is transplanted into the GI tract of a sick person, the procedure was found to be more than 90 percent successful—an astonishing and unprecedented cure rate.

The procedure typically requires invasive and uncomfortable colonoscopies, enemas, and even nasal tubes to deliver the healthy stool. However, researchers have recently refined the delivery method to frozen pills, and have found them to be just as safe and effective as traditional transplant techniques. Oh, the sweet smell of progress!

The Immune Tuner

Autoimmunity—when a person's immune system attacks parts of their own body—is the defining feature of many common diseases, including celiac disease, MS, type 1 diabetes, and Hashimoto's disease, to name a few. Why does autoimmunity develop, and why is it seemingly on the rise? Are we meant to have immune systems that harm our bodies and brains in acts of friendly fire, or is this one more aspect of our biology succumbing to the pitfalls of modern life? To understand how our diets and lifestyles may contribute to a confused immune system—and thus autoimmunity—it helps to understand how this dynamic system is "trained" throughout your life.

If you were to picture a cross section of a tunnel, you'd get a pretty good sense of the anatomy of the large intestine. The innermost tissue is the epithelium. This one-cell-layer-thick barrier acts as the division between the interior of your colon—known as the *lumen*—and your circulation. Because its contents are not part of your body (similar to the air that fills your lungs), the lumen is actually thought of by scientists as being part of the host's environment. In fact, the gut is actually your largest interface with the environment—much larger than your skin. If you were to remove your entire digestive tube from your body, uncoil it, and sprawl it out on the ground, it would take up the square footage of a small studio apartment.

For this reason, the vast majority of our body's immune cells are primed to focus on what's happening in our digestive systems. Though this may seem counterintuitive, or even a misuse of resources in today's world of packaged foods and triple-washed produce, it makes sense: for the vast majority of our time on Earth, well before any semblance of a modern food system, our food was *dirty*. We didn't have supermarkets full of the freshest (and most attractive) produce at our convenience, and we certainly didn't coevolve with the plethora of antibacterial soaps or "vegetable washes" available today, promising hospital-grade sterility to our every last bite.

For our Paleolithic ancestors, the potential for swallowing a pathogen—a microbe that could infect and possibly kill us—was great. This put immense pressure on our species early on to ensure that we could mount an agile and formidable immune response should such a confrontation occur. But our intestines are *loaded* with outside bugs—is there a war raging in our bellies that we aren't privy to?

Not quite. A healthy immune system must work like the highly trained security personnel at a sports arena, deftly surveying thousands of ticket-holding attendees without breaking a sweat. These guards don't question every quirky-looking stadium visitor they see—when they're well trained, they can spot signs far in advance if a person is likely to step out of line. Like the visitors at a stadium, our microbial residents hone our guards' skills, helping them adapt to the constantly changing environment—so that when an unfriendly visitor *does* come along, it can be easily spotted. Our gut—and its inhabitants—therefore serves as a sort of "training camp" for our body's immune system.

When our immune system is not up to par, it not only becomes less effective at spotting invaders, but sometimes mistakenly attacks the body's own cells. This is because a diverse gut population not only teaches our immune system guards who to watch out for, it also teaches them the importance of tolerance. In a healthy gut there may be hundreds or thousands of different species present at any given time, and a healthy immune system benefits from this plurality of voices. In fact, this is partly how probiotics are believed to work: they consist of species that aren't normally resident in our microbiomes, flowing through us to make sure our guards aren't snoozing on the job.

Issues like allergies and autoimmunity develop when the immune system makes a mistake and attacks its own host, and the microbiome has become a focal point for scientists working to figure out why this happens. It has been proposed that our immune systems have become dysfunctional for a number of reasons, including our overly hygienic lives, antibiotic overuse, fiber inadequacy, and birthing practices that place the developing microbiome as an afterthought. Any one of these factors, it is believed, may lead to stadium guards that aren't as well trained—and thus to higher rates of autoimmunity.

CLEANING OURSELVES SICK?

Something other than our food supply has changed over the past couple of decades—we've become more sterile. But in our preoccupation with eliminating any possibility of a stray virus or pathogenic bacteria, we've essentially wiped clean many of the more positive interactions that we would have had with rogue bacteria. Such interactions help to *train* the adaptive immune system, which, after all, was shaped by natural selection under just these circumstances.

Research shows that while exposure to pathogens (and the rate of infections) has decreased, rates of both autoimmune and allergic diseases have *increased*. The idea that these two statistics are causally linked is the basis for the "hygiene hypothesis." The theory is this: some infectious agents—notably those that coevolved with us—protect us from immune-related disorders. Today, the absence of these pathogens results in weakened immune systems, leaving them vulnerable to confusion and setting the stage for type 1 diabetes, multiple sclerosis, celiac disease, and others.[14]

With diabetes, obesity, and even Alzheimer's disease all characterized by chronically heightened inflammation—aka an immune system gone awry—it isn't a leap to suggest that our overly sterile lives may be to blame here as well. In fact, recent research has explored the very link between national hygiene and Alzheimer's disease incidence. Using public sanitation and access to clean drinking water as metrics, researchers revealed a striking relationship: countries with greater levels of hygiene had increased incidence of Alzheimer's disease, in a perfect linear correlation.

Gluten provides the perfect illustration of how a confused immune system could lead to an autoimmune response, as it does for a significant portion of the population. Gliadin, one of the main proteins in gluten, looks a lot like a microbe to our immune cells. When it is present in the gut, our immune system sends out antibodies in a manhunt for antigens—physical attributes that our stadium security guards are trained to look out for. The problem is, antigens on foreign substances (like gliadin) can look uncomfortably similar to the markings on our own cells. This is called *molecular mimicry* and may be an attempt by pathogens to better fit into the host environment—because even pathogens have a drive to survive! This means that when the body's immune system creates antibodies to fight antigens, our own tissues can fall under friendly fire.

This can often occur to a family of enzymes called *transglutaminases*. Present throughout the body, transglutaminases are important for keeping us healthy, and their dysfunction has been implicated in Alzheimer's disease, Parkinson's disease, and ALS.[15] They're also found in particularly high concentration in the thyroid, which gets attacked in autoimmune thyroid conditions such as Hashimoto's disease and Graves' disease. Unfortunately, transglutaminase enzymes have molecular markings very similar to gliadin antigens. In a susceptible person, eating gluten may lead the body to attack not only gliadin, but also the transglutaminase enzymes.

While it can't be said that gluten is the cause of a mutinous immune system for every person, a recent study found the prevalence of celiac disease in patients with autoimmune thyroid disease was two- to fivefold compared to healthy controls.[16] In fact, other autoimmune conditions (including type 1 diabetes and multiple sclerosis) occur together with celiac disease more frequently than with any other autoimmune condition, suggesting that an unhealthy gut is a mediator of this wide range of seemingly unconnected diseases. Any one of these conditions may be a sign that the brain is under threat of attack: people who have autoimmune disorders are more likely to develop dementia, recent research shows.[17] Keep in mind that these diseases manifest after many months or years, often without obvious symptoms. And, for many patients with overlapping thyroid and celiac disease, gastrointestinal symptoms are absent, marking one of the rare instances where "going with your gut" may lead you astray.[18]

Preventing or halting this immune breakdown is not achieved simply by going on a gluten-free diet. It's important to add something back to the diet that is sorely missing from our modern plates: fiber. Fiber directly protects us against immune confusion in part because SCFAs like butyrate increase the production and development of *regulatory T-cells* in the colon. These cells, also called *T-regs*, are a type of immune cell that helps ensure a healthy and appropriate inflammatory response by suppressing the responses of other immune cells, including those that promote inflammation.[19] Think of them as the managers on the security force that keep the more pugnacious junior guards under control. They are key players in helping your body better distinguish between itself and everything else. If that critical ability breaks down, your immune system might end up attacking your own body, and voilà—autoimmunity develops.

Protecting Our Brains from What's in Our Guts

As I've mentioned earlier, the colon is where the majority of bacteria in your gastrointestinal tract reside, giving the cells that line it two important functions: serving as a blockade against pathogens and bacteria that don't belong in your circulation, while also allowing for the absorption of fluids and any remaining nutrients that haven't been absorbed by the small intestine. This physical barrier forms part of the body's *innate* immune system.

The innate immune system plays a major role in mediating inflammation and autoimmunity. It helps keep the microbiome and our immune cells (our *adaptive* immune system) separate from each other, regulating host-microbial interactions and maintaining appropriate immune function on a constant basis. In our stadium metaphor, this allows the game to go on as planned, ensuring a nice day for everyone there. The guards can safely perform their jobs, the fans can eat their hot dogs and cheer for their respective teams, and the players can compete, allowing them to earn millions of dollars in endorsement money. Physical barriers help make all this possible.

The cells of the epithelium—your intestinal lining—are held together by tight junctions that can open and close like the drawbridge on a castle. Thankfully, they are closed most of the time. However, exposure to potentially dangerous bacteria, especially in the small intestine, can cause the junctions to loosen, drawing water and immune cells into the gut lumen. This usually results in diarrhea to flush out the troublemaker—a critical defense response during acute infection.[20] Unfortunately, certain aspects of modern life can also cause our gut barrier to be more porous and allow for *retrograde transport*, or the transport of gut contents deep into the gut lining. This leads to considerable consequences and possibly initiates the "molecular mimicry" that is thought to result in autoimmunity.

One possible instigator of undue permeability to the gut lining is gluten, the protein found in wheat, rye, barley, and many packaged foods. Gluten is unique among proteins that we consume because unlike, say, the protein we get from eating a chicken breast, gluten is not completely digested by humans. Protein from most sources will separate into its constituent amino acids during digestion, but gluten breaks down only into large fragments called *peptides*. These fragments have been found to stimulate a more permeable gut in humans, triggering a welcoming from the innate immune system more similar to that of a bacterial invader than that of your usual dietary protein.

At the center of this response is another protein called *zonulin*, which is produced in the gut whenever gluten is present.[21] Zonulin acts as a cellular gatekeeper of sorts, regulating the integrity of the tight junctions in between epithelial cells. Where there is zonulin, there is permeability. (Dr. Alessio Fasano, founder of Mass General's Center for Celiac Research and an internationally recognized expert in celiac disease, is credited with having discovered this important mediator of gut permeability.) This "hyperpermeability" can occur in anyone but is exaggerated in those with celiac disease. For this population, gluten evokes an overt autoimmune response, causing the lining of the small intestine to become damaged over time.

One of the dangers of a more permeable gut is that it allows bacterial *endotoxin* (also known as *lipopolysaccharide*, or LPS) to cross over into circulation. As I've mentioned in previous chapters, LPS is a molecule that makes up part of the membrane of certain bacteria that normally live within the safe harbor of our large intestine. When leaked into circulation, endotoxin sets off an acute pro-inflammatory response, signaling a systemic bacterial invasion. LPS exposure is directly related to pro-inflammatory cytokine production and an increase in oxidative stress, wreaking havoc on a wide range of bodily systems—including your brain.

When animals are inflamed, usually from infection, they exhibit strange behavioral changes, displaying symptoms of lethargy, depression, anxiety, and reduced grooming. They retract from the herd and become more sedentary, a means of reserving the body's energy for healing and isolating them from the healthy. This is not a phenomenon exclusive to farm animals—humans react similarly. They become irritable, lose interest in food and socializing, and have trouble focusing and even remembering recent events.[22] These are called *sickness behaviors*, and it's a phenomenon well known to farmers, zookeepers, and scientists. Psychologists believe this to be a motivational state—an adaptive strategy on the part of our biology to aid in survival.

Major depression may be an extreme form of sickness behavior. Depression is well known to be more common in those with inflammatory conditions like heart disease, arthritis, diabetes, and cancer. On the surface, these conditions have nothing to do with the brain, but the volume of inflammatory markers in the blood and the risk for depression correlate in lockstep—the higher the levels of these markers, the more severe the depression.[23] This groundbreaking new view of depression, a condition affecting more than 350 million people globally, has challenged the preexisting paradigm for treatment and given rise to a whole new theory of its origin: the inflammatory cytokine model of depression.[24] And what is often injected to induce such a state in lab animals by scientists studying depression and other consequences of inflammation? Bacterial LPS.

Zonulin, the protein that triggers increased permeability, is also capable of altering the tight junctions at the blood-brain barrier, which is another layer of specialized epithelial cells. This is significant because the breakdown of the blood-brain barrier has been implicated early on in the development of Alzheimer's disease. Not surprisingly, a gluten-free diet reduces both zonulin levels and gut permeability, and may maintain the brain's protective barrier as well.[25]

So, if you don't have celiac disease or a wheat allergy, might cutting wheat out of your diet help your brain work better? Researchers at Columbia University recently asked this very question, and studied patients who did not have celiac disease or a confirmed traditional wheat allergy. However, they did have symptoms such as fatigue and cognitive difficulties after eating wheat. The researchers placed the subjects on a wheat-, rye-, and barley-free diet, and after six months, signs of immune activation and intestinal cell damage had disappeared. This was associated with significant improvement in both gastrointestinal symptoms and cognitive function, as reported by the patients via detailed questionnaires.[26] While the medical community has been debating and disputing the very existence of wheat sensitivity, this exciting research is among the first to validate nonceliac wheat reactivity with objective measurements.

HOW DO PROBIOTICS WORK?

Though it may seem like probiotic drinks, supplements, and even probiotic-infused foods are all the rage today, probiotics are by no means a new phenomenon. We've been fermenting foods and harnessing the power of live bacteria to preserve perishables for millennia. The earliest record of fermentation dates back more than eight thousand years—and nearly every civilization since has included at least one fermented food in its culinary heritage. In Japan there's *natto*, in Korea *kimchi*, Germany loves its *sauerkraut* (as do I), and modern yogurt, now ubiquitous, holds on to its original Turkish name!

While many believe probiotics are helpful because they take up domicile in the gut, the truth is that the vast majority of probiotics we consume are just transient visitors, offering friendly communication with our more permanent residents and our immune cells.[27] Our immune systems work best in a state of happy harmony with our microbiota, and probiotics seem to foster this connection, essentially "tuning" the immune system on their journey south. Probiotics can also reinforce the precious gut barrier, "plastering up" any leaks in the tight junctions between our gut epithelial cells. This can prevent compounds like *endotoxin* from leaking into circulation, a major instigator of systemic inflammation. Both of these functions together help explain the anti-inflammatory effects demonstrated by probiotics. The downstream benefits are innumerable and support the observation that people who eat more fermented foods tend to have better health and quality of life.

Just remember: taking a probiotic supplement alone will never fix the damage caused by a poor diet, but the current data does show that consuming probiotic-rich foods like kimchi, kombucha, and kefir can enhance the effects of a high-fiber, low-carbohydrate diet as described in this book. Though supplementation is not often necessary, it can't hurt, and it might help. In chapter 12 I will explain how to pick a high-quality probiotic supplement, should you chose to go that route.

Before we close out this section on our glorious gut barrier, it's important to note that gluten isn't the only potential instigator of increased permeability. Here are other factors that can lead to a more porous gut:

- **Alcohol consumption.** In healthy nonalcoholics, a single vodka binge dramatically increases endotoxin and proinflammatory cytokines in the blood.[28] That's because alcohol has been shown to inflame and produce a more permeable gut, explaining in part the harm that chronic alcohol consumption confers on the liver and other organs.[29]

- **Fructose.** When fructose is removed from the matrix of fiber and phytochemicals that is normally found in whole fruit, it can increase intestinal permeability. High-fructose corn syrup or agave syrup, which are widely used to sweeten commercial beverages, may be particularly bad news.
- **Chronic stress.** Public speaking (a common stressor for many) has been shown to momentarily induce gut permeability in humans, pointing to a new mechanism through which chronic stress can harm our health.
- **Excessive exercise.** Endurance athletes may experience gut permeability due to the stress of sustained aerobic training.[30] In chapter 10 I'll share new research on exercise that renders those long, grueling cardio sessions totally unnecessary.
- **Fat, when consumed together with sugar.** In animal models, high-fat diets (which often include sugar) have been found to induce "leaky gut" and inflammation.[31]
- **Processed food additives.** More on this in a moment.

Any of these stimuli can facilitate the leakage of endotoxin into circulation, even if you're on a gluten-free diet. Conversely, various plant compounds such as quercetin (a polyphenol found in onions, capers, blueberries, and tea) as well as the amino acid L-glutamine have been found to reduce gut permeability and promote better function of the gut lining.[32] And fiber, that miraculous and underrated nutrient, may be the most important of all, through its effect on an important albeit slimy structure called the *mucosa*.

Our Marvelous Mucosa

Thankfully, the epithelial cell layer doesn't have to defend against our constant daily onslaught of toxic substances and microbes on its own.

In between the epithelium and the trillions of microbial cells of the microbiome, there is a dynamic matrix of mucus known as the *mucosa*, which comprises a form of carbohydrate called *mucin*. This mucus is produced by the cells of the epithelium and is essentially where the rubber hits the road for the microbiome—not only does it provide a soft hammock upon which the bacteria luxuriate, but the mucus itself acts as a "demilitarized zone"—a layer of protection for the epithelial cells.

Keeping the mucus layer healthy and robust is a major mechanism through which we can minimize inflammation in the body and, likely, the brain. While the science around this is new and evolving, one surefire strategy is to ensure a steady flow of dietary prebiotic fiber. This fiber feeds the microbes that supply butyrate, which actually feeds the cells that create the mucus, thus reinforcing its protective abilities.[33] Conversely, a low-fiber diet starves our gut bacteria, forcing them to actually *consume* the mucosal layer out of desperation.

"BUT I'M NOT A MOUSE!"

Whenever one discusses early-stage research, inevitably conducted in animals prior to humans or when human studies are unethical or impractical, there is always the question of how well it would apply to humans in the real world. A perfect example of this paradox is that Alzheimer's disease has been cured in mice many times over, and yet the results have never translated to human trials. The truth is that mice don't get Alzheimer's disease in the wild, and scientists in these cases are working with an artificially induced Alzheimer's disease *model*—an imperfect simulation.

On the other hand, basic cellular mechanisms are highly conserved by evolution, meaning that they differ minimally across species. The more basic the process, the further away from humans we can go and still see an accurate result. We can study cell division by looking at yeast, for example. We can study the way a human neuron works by looking at the nearly identical giant neuron of a squid. And we can study the development of the fetal brain by looking at fruit flies. These processes are so important that they differ very little from species to species, thus enhancing our confidence in the translation of the results.

As we delve into the study of the intestinal lining, we can infer much from animal studies, as the cells lining and surrounding our guts are very similar in all mammals.[34] Is it certain that industrial chemicals have the same effect in humans as they do in mice? It will be years before the human science is settled, but we must make decisions about the foods we ingest *today*.

Gluten is a good example of a protein that some people may be able to tolerate in small, infrequent doses, but that irritates the gut lining in the context of the Western diet, which is low in fiber and high in bread, pasta, and packaged products. Packaged foods are full of emulsifying agents, which are used to create tasty mixtures of otherwise insoluble foods and ensure a smooth texture. These are commonly found in salad dressing, ice cream, nut milk, coffee creamer, and other processed foods. In animal studies, adding even a small amount of emulsifiers to the diet caused a profound change in the gut microbiota, eroding the mucosa and reducing the average distance between gut bacteria and intestinal cells by more than half.

It may be that a "one-two punch" is required to initiate an inflammatory process—first the erosion of the gut's protective layer, and then a reaction within the gut lining. When the mucosa is compromised, gut bacteria—both the beneficial, butyrate-producing ones as well as pathogens—are able to infiltrate our gut barrier. This can lead to inflammation of the gut as the normal bacterial inhabitants breach the mucosa and come too close for comfort to our own immune system. In the emulsifier study I just mentioned, this is exactly what happened to the animals.[35]

The critical new insight here is that it may not simply be certain proteins (e.g., gluten, or lectins, another class of plant protein that has caused a stir lately) that cause this gut breakdown and inflammation in so many people. Rather, the very act of consuming industrially processed foods—those that have been stripped of fiber and made with agents like emulsifiers to create a smooth mouthfeel—can independently alter our microbiomes, strip our mucus linings, and make more of us vulnerable to the effects of those proteins.

We Breed What We Feed

The gut microbiome is much like an actual city in that there are at least a thousand different species living in an immensely complex and highly competitive environment. There are beneficial, SCFA- and butyrate-producing bacterial species, and there are problematic bacterial species, including potential pathogens (bacteria that might actually make you sick) that are kept under control by the community at large.

THE POTENT POWER OF PROBIOTICS

Ready for a paradigm shift? Some very interesting studies have emerged recently highlighting the value that probiotics—foods or supplements rich in live bacteria—may hold for those of us suffering from depression, anxiety, and even dementia.

In a small study from Holland's Leiden Institute for Brain and Cognition, women who took a probiotic supplement designed to enhance gut bacterial diversity experienced less reactivity to sad thoughts than those who took a placebo. Resilience to sad thoughts is a sign of strong mental health. For example, in depressed people, a sad stimulus can turn an otherwise spotless sky into an overcast day, whereas someone with a healthy mood can simply observe the sad thought and move on, without significant cloud formation.

Can consuming more fermented foods, like kombucha, yogurt, sauerkraut, and kimchi help our anxiety? Maybe, according to another study that found that students who consumed more of these types of foods had less social anxiety. The effect was especially strong in those who had neurotic personalities. "It is likely that the probiotics in the fermented foods are favorably changing the environment in the gut, and changes in the gut in turn influence social anxiety," wrote one of the authors in the College of William and Mary and University of Maryland study.

Groundbreaking research out of Iran suggests that probiotics may even boost cognitive function in one particularly desperate group: patients with advanced Alzheimer's. The researchers put severely demented patients on a twelve-week, high-dose cocktail of *Lactobacillus* and *Bifidobacterium* (two common strains of probiotics) and found that, compared to the control group who received only a placebo, the probiotic group improved on a test of cognitive function by an astonishing 30 percent. Though the effect would need to be replicated with a larger sample size, these preliminary results are certainly cause for hope.

Scientists have truly just scratched the surface of what will be a fascinating decade to come in microbiome research as the broad reach of probiotics comes into view. Certain strains may help fight off certain cancers, boost heart health, enhance brain neurogenesis, and even alter mood states—the latter paving way for "psychobiotics" (more on this in chapter 8).[36]

The two reigning bacterial families found in the large intestine are the Bacteroidetes and Firmicutes. They're sort of like the Montagues and the Capulets, if *Romeo and Juliet* took place in your large intestine. Though there is currently no consensus on what the "perfect" microbial composition looks like, there are correlates that scientists can draw by observing the microbial signature of various populations with different health profiles. For example, some research has suggested that overweight people have more Firmicutes than Bacteroidetes (or Capulets than Montagues, in our Shakespeare analogy). At this point, it is unknown whether this or any other signature is causally related to, or merely reflective of, the health of its human host. However, animal studies using fecal microbial transplants are paving a road toward greater clarity. With this method, we're able to answer the question: can we change aspects of an animal's health and appearance by changing its microbiome?

In one such example, scientists wanted to see what would happen if they transplanted the microbiomes of obese mice with insulin resistance into the digestive tracts of lean mice. As if by magic, the lean mice, when given the obese mice's microbes, began to gain weight, displaying the same metabolic dysfunction as their obese counterparts.[37] While humans are more complex than mice, this study does suggest that microbes, in many ways, call the shots—at least where our weight is concerned. But what about our mental health and cognition?

For the first time, groundbreaking research illustrated a link between brain structure and function and gut bacteria in healthy humans. In this UCLA study, healthy women had their microbiomes sequenced and brains scanned and were given a test that assesses risk for depression. Those women with a higher proportion of *Prevotella*, a type of bacteria, in their guts had heightened connectivity between emotional and sensory brain regions while having smaller and less active memory centers.[38] When they were shown negative imagery, these women seemed to experience stronger emotions, as if they were distressed. On the other hand, women with a higher proportion of *Bacteroides*, another common type of gut bacteria, were much less likely to experience negative emotions when shown the same images. Structurally, their memory centers were larger, and they also had more volume in their prefrontal cortexes, which is the hub of executive function. It seemed as if the women with less *Prevotella* and more *Bacteroides* were emotionally stronger and more resilient.

Were the bacteria affecting the women's brains, or were the women's brains somehow altering the mixture of bacteria in the gut? No one knows. However, as they have done with metabolism and weight, scientists have been able to alter mouse behavior and what could be interpreted as mouse mental health just by tinkering with their microbiomes, suggesting that the types of bacteria present in the gut do play a role in brain function.[39]

As I've mentioned, the optimal gut makeup is a puzzle that is a long way from being solved, and it is likely to be different for you than it is for me. It is interesting to note, however, that people who consume carbohydrate-rich, grain-based diets tend to have higher proportions of *Prevotella* bacteria residing in their guts.[40]

Many scientists in the field seem to agree that the best way to ensure that the beneficial bacteria maintain a competitive edge in the tough and constantly changing colonic environment is to consume a diet that is rich in fiber and plant nutrients such as polyphenols, and to avoid sugar and refined carbohydrates. This pattern will directly benefit the helpful microbiota and will starve out the pathogens, making it difficult for more malevolent species to gain a foothold in the rough-and-tumble gut ecosystem. As we await further clarity, trading a grain-based diet in for one built on prebiotic fiber–rich vegetables seems like a safe bet toward shifting your microbiome (and mood) to a healthier state.

FAQ: But whole grains contain fiber—shouldn't I eat more of them?

A: Whole grains contain very small amounts of prebiotic fiber. The fiber content of grains is mainly **insoluble** fiber, and where your microbiome is concerned, all fiber is not created equal. Insoluble fiber is not prebiotic and is unable to be metabolized by gut bacteria (it's basically sawdust). Grains also provide a large amount of starch, which is essentially pure glucose. Given the low amount of prebiotic fiber and the high amount of glucose, whole grains are probably not the best way of attaining your daily dose of fiber.

Diversity Rules

As I've mentioned, our immune systems benefit from a plurality of bacterial voices, and yet diversity is another area where our modern microbiomes are lacking. Many studies comparing the gut microbiomes of Western city dwellers to rural villagers and hunter-gatherers who eat more plants (and thus more fiber) have shown the striking loss of diversity driven by such modernization. By ensuring that your diet is rich in a mix of different *types* of fiber, as each type of fiber feeds different species of bacteria, you directly foster gut microbial diversity—a feature that researchers, even at this nascent state in microbiome study, agree is key to the health of the host. In fact, research has shown that fiber alone can dramatically increase or decrease gut microbial diversity, a quality that you might even be able to pass along to your kids.[41] Here are some other ways to maximize your gut microbial diversity:

- **Avoid antibacterial soaps and hand sanitizers.** Use only when absolutely necessary, such as when visiting areas of high pathogen exposure risk, like hospitals.
- **Embrace nature.** Spend more time outdoors, in parks, camping, or hiking.
- **Consume filtered water.** The use of chlorine to eliminate outbreaks of waterborne pathogens in developing nations is

a great thing, but many first-world water supplies tend to be overtreated with chlorine.

- **Shower less.** Or use soap more sparingly, perhaps only every other shower. The resulting increase in mating scent molecules called *pheromones* may even help your dating life. Shampoo once or twice a week at most—there's no reason to shampoo every day!
- **Buy organic produce whenever possible.** Organic produce will be richer in antioxidant polyphenols, which support butyrate-producing bacteria as well as a healthy mucosa.[42]
- **Avoid taking broad-spectrum antibiotics unless absolutely necessary.** Antibiotics can save lives when appropriate—this is an undeniable truth. However, 30 percent of antibiotics prescribed in the United States are completely unnecessary according to recent research, and they can devastate the microbial ecosystem.[43] This can make room for opportunistic pathogens like *C. difficile* to take over instead.
- **Adopt a pet.** There are millions of homeless animals in shelters all around the United States that would be happy to help you increase your microbial diversity. Women who have a dog in their homes when pregnant are less likely to have children with allergies, and kids who grow up with dogs are 15 percent less likely to develop asthma.[44] Living with a dog is one of the top ways to increase the microbial diversity of the home and in the gut.
- **Slow down.** Digestion takes place when you are relaxed, hence the term "rest and digest." Eating on the go can set off a cascade of stress response mechanisms in the body that compromise digestion, not only impairing your absorption of nutrients but affecting your bacterial friends' access to them as well.

A Bright Future

The more we learn about the gut, the more we're coming to understand the role it may potentially play in the development of various diseases. At the same time, we're starting to see how tending to it might help treat those conditions as well.

Many neurological and even psychiatric conditions are associated with intestinal inflammation, and are preceded by symptoms in the gut. Autism spectrum disorder (ASD) has been closely linked with gut inflammation, which coincides with inflammation in the brain.[45] Many autistic children have intestinal problems such as inflammatory bowel disease and an excessively permeable gut lining. In a test of intestinal permeability (called the *lactulose-mannitol test*), 37 percent of ASD children tested positive, compared to less than 5 percent of control group kids—that's a sevenfold increased incidence. This effect size certainly suggests a possible causal link, with either intestinal permeability causing autistic behavior, autism causing increased intestinal permeability, or some third factor, such as an environmental exposure, causing both.

At the other end of the age spectrum, Parkinson's disease, a neurodegenerative condition, has also been strongly linked to gut health. One of the earliest, and often overlooked, signs of the disease is constipation. Though scientists are still working to understand this, a significant clue was revealed in a recent study involving fifteen thousand patients with severed vagus nerves. The vagus nerve sends messages from the GI tract directly to the brain, and only half of these patients with severed nerves developed Parkinson's disease over twenty years, compared to rates seen in the general population. This is strong evidence that Parkinson's disease might in fact begin in the GI tract and travel up the vagus nerve to the brain.[46]

Fecal microbiota transplantation—the transplantation of healthy stool that is up to 60 percent bacteria by weight—is an exciting development because it provides an opportunity to press "reset" on the gut microbiome. Today, this procedure involves transplanting intact stool, which can contain thousands of types of bacteria. Scientists do not yet know exactly which ones hold the curative powers, but in the future, the potential for more selective bacterial interventions on a whole host of conditions will undoubtedly emerge.

It's important to remember that the gut microbiota is but one element of our human-microbial symbiosis. Emerging areas of research include the oral and sinus microbiomes. Poor oral health has been long linked to a number of systemic illnesses, including stroke, diabetes, cardiovascular disease, and dementia.[47] In a paper published in *PLOS ONE*, researchers found that patients with mild to moderate dementia with periodontitis—gum inflammation—had a sixfold increase in the rate of cognitive decline six months later.[48] Should we be swishing with antibiotic mouthwash, essentially nuking the friendlier bacteria populations in our oral microbiome along with those that may be more harmful? Do the same opportunistic species found in the gut lurk below the gumline, waiting for the right moment to stage a coup? These are questions that future research will undoubtedly need to address.

The sinus (or rhinosinal) microbiome may be of particular relevance to the brain. The sinus cavity provides direct access to the brain via its rich vascular bed of highly permeable capillaries. What does this mean for the microbial chemical factory that inhabits that area? Recent research out of Harvard suggests that amyloid plaque (the kind that builds up in Alzheimer's disease) may be, for some, a response to a brain-microbial infection. This positions the nasal microbiome as an exciting candidate for exploration in the coming years. What mixture of microbes provides the least competitive advantage for incoming troublemakers? Are probiotic nasal sprays a cognitive-boosting treatment of the future? Is it a coincidence that sense of smell is the first sense to be affected by cognitive decline? I, for one, am excited to track the developing science on the horizon.

FIELD NOTES

- A healthy gut becomes a butyrate factory, transforming dietary fiber to one of the most important quenchers of inflammation.
- Butyrate has been shown to boost BDNF, the brain's ultimate fertilizer.
- Autoimmunity (when the host's immune system attacks the cells of the host) can be instigated by gluten for many people. The typical low-fiber American diet (rich in emulsifiers) may exacerbate the threat posed by gluten.
- Gut bacterial diversity is important for "training" a healthy immune system—a feature modern life has greatly reduced.

GENIUS FOOD #7

DARK LEAFY GREENS

Vegetables are your brain's best friend. There are no ifs, ands, or buts about it, especially when we're talking about the nonstarchy varieties including spinach and romaine lettuce, and the cruciferous veggies cabbage, kale, mustard greens, arugula, and bok choy. These dark leafy greens are low in sugar and packed with vitamins, minerals, and other phytonutrients that the brain desperately needs to function properly.

One of the nutrients that dark leafy greens are full of is the vitamin folate. In fact, the word *folate* comes from the Latin word for "foliage," making it pretty easy to remember how to get more of it: eat leaves! Known mostly for its ability to prevent neural tube birth defects, folate is an essential ingredient in your body's methylation cycle. This cycle occurs on a constant basis throughout the body and is critical for both detoxification and getting your genes to do their proper jobs.

Another important nutrient found in greens is magnesium. Magnesium is known as a "macromineral," because we need to get a relatively large amount of it from our food for optimal health and performance (other macrominerals include sodium, potassium, and calcium). Nearly three hundred enzymes rely on magnesium, making it pretty popular around the body. These enzymes are tasked with helping you generate energy and repairing damaged DNA, which is the underlying cause of cancer and aging, and even plays a role in Alzheimer's disease. Sadly, magnesium consumption is inadequate for 50 percent of the population. But lucky for us, anything green is usually a good source of magnesium, as this mineral is found at the center of the chlorophyll molecule (which gives plants their green pigmentation). Perhaps this is why a recent study has shown that people who ate just two servings of dark leafy greens a day had brains that looked eleven years younger on scans!

Dark leafy greens also provide an undeniable benefit to us by way of the fiber that they contain. In chapter 7 you learned all about the gut microbiome and its collective ability to produce *short-chain fatty acids* like butyrate—a powerful inflammation inhibitor. The number one way to feed these microbes (and in turn extract butyrate for ourselves) is to increase vegetable consumption, which ensures a diverse and ample pipeline of fermentable, prebiotic fibers for our microbe friends. Leafy greens even contain a newly discovered sulfur-bound sugar molecule called *sulfoquinovose* (try to say that three times fast) that directly feeds healthy gut bacteria.

Overall, consumption of vegetables—and dark leafy greens in particular—benefits both brain and body, and is even inversely related to dementia risk and various biomarkers for aging.

How to use: Eat one huge "fatty salad" daily, which is a salad filled with organic dark leafy greens like kale, arugula, romaine lettuce, or spinach, and doused with extra-virgin olive oil. Avoid nutrient-poor varieties like iceberg lettuce, which is essentially just water and fiber. There will be more "fatty salad" options in the recipe section (pages 323 to 342).

CHAPTER 8

YOUR BRAIN'S CHEMICAL SWITCHBOARD

My first attempt at decoding the word *neurotransmitter* (and the many drugs that affect neurotransmission) came during that crystalline moment immediately following my mom's diagnosis, when we sat in the rental car in a parking lot outside the Cleveland Clinic. I was trying to sound out the drug names on the various pill bottles in our possession after our visit to the pharmacy.

The brand names were bizarre clusters of vaguely word-like phonemes—consonant-vowel-consonant followed by vowel-consonant-vowel combined in mellifluous, pleasing sequences. They looked like they *could* be words, *should* be words, as though one were reading the pages of a book written in English in a parallel dimension. *Na-men-da. Sin-e-met. Ari-cept.* How naturally they might slip into our daily conversations.

"Bro, what are you doing tonight?"

"Namenda."

"Me neither. Let's go to the Sinemet."

"I don't know if Aricept that."

The generic, scientific names were obviously less tailor-made for a television audience, leaving an anxious ambiguity as I rolled them around on my tongue: *memantine*, *levodopa/carbidopa*, *donepezil.* Was that *dawn-EH-pazeel*? Or *DONNA-pezel*? Where does the accent go? I wondered. I would settle on one pronunciation, fairly confident I had nailed it, until a doctor would use a completely nonintuitive pronunciation. Wow, the things you learn in medical school! I would then go to another doctor's office, ready to impress him or her with my proper pronunciation, only to have *that* doctor smirk, confidently asserting that this third variant of *donepezil* ("Everyone knows that the first *e* is silent!") was in fact the authoritative version.

Pronunciation aside, what do these drugs actually do? These quirky-sounding compounds work by altering levels of neurotransmitters. Dementia drugs are not the only compounds that do this—many prescription drugs, from antidepressants to ADHD medications to drugs that reduce anxiety, tinker with levels of these important chemical messengers. While these types of drugs are among the top-selling pharmaceuticals in the world, other compounds that humans have gravitated to across cultures also work similarly—coffee, alcohol, cocaine, MDMA, and even sunlight all make us feel a certain way because of their impact on how neurotransmitters work.

The notion that our brains don't work the way we want them to because of unbalanced levels of neurotransmitters has come to be known as the "chemical imbalance" theory. This theory is most commonly associated with depression, for which it states that feeling blue is caused by low levels of serotonin in your brain. But new research suggests that many common brain problems are not caused by deficits of neurotransmitters, but rather by neurotransmitters that are unable to work the way they ought to, due to an induced or underlying dysfunction. Similarly, dementia is not *caused* by low acetylcholine, a neurotransmitter involved in memory; acetylcholine is low because the neurons that produce it are, in many cases, slowly dying.

This is why such drugs have no "disease-modifying" ability—meaning, they do nothing to solve the underlying problems that create the package of symptoms that we see as "dementia." They act merely as Band-Aids. Attention deficits, memory loss, and depressed mood may all be manifestations of underlying problems, and pharmaceuticals continually come up short.

HOW NEUROTRANSMITTERS WORK

For a microscopic system, neurotransmitter function is an incredibly elegant design. Some of the neurotransmitter gets released by a neuron. This neuron is called the *presynaptic* cell because it initiates the message and thus comes before the synapse. The neurotransmitter then moves into the synaptic *cleft*, which is the gap between neurons. There, molecules of neurotransmitter cross the gap to meet a receptor on the receiving, or *postsynaptic*, neuron. The leftover neurotransmitter is either taken back up by the presynaptic cell, called *reuptake*, or degraded by enzymes. Under normal conditions, this postsynaptic "cleanup" is done to prevent excessive stimulation of the postsynaptic cell, but in certain cases it may be manipulated by drugs to various effect. For example, inhibiting reuptake is the mechanism by which certain antidepressant drugs increase availability of the neurotransmitter serotonin. Alternately, drugs that aim to boost acetylcholine in the brain, another important neurotransmitter, work by preventing its enzymatic degradation.

This chapter will explore how to keep your neurotransmitters functioning optimally by helping you recreate the conditions that they were designed to work under. Whether you suffer from poor mood or memory, stress, or lack of focus, this section will help you to better understand how to maximize your quality of life, cognitive function, and brain health through the brain's main means of communication. Because a healthier brain enhances our experience of the world, it allows us to be the truest and most expressed versions of ourselves, capable of feeling, learning, loving, and connecting in ways that make life worth living.

Glutamate/GABA: The Yin and Yang Neurotransmitters

Ancient Chinese philosophy describes life in terms of the inhibitory (yin) and the excitatory (yang) coexisting in perfect harmony. Without knowing it, these ancient philosophers seem to have stumbled upon a rudimentary description of our two most fundamental neurotransmitters thousands of years prior to the invention of the scientific method!

GABA is the chief inhibitory neurotransmitter in the brain, used in 30 to 40 percent of synapses brain-wide. Associated with a calming effect that has been dubbed "nature's Valium," it works to counterbalance glutamate, the brain's chief excitatory neurotransmitter—the *yang* to GABA's *yin*. GABA and glutamate make up the brain's most plentiful neurotransmitters and are involved in the regulation of vigilance, anxiety, muscle tension, and memory functions.[1]

Glutamate

Used by more than half of all neurons, glutamate is GABA's precursor and increases the brain's overall level of excitation. Glutamate is normally involved in learning, memory, and synaptogenesis (the creation of new connections between neurons).[2] We've already covered a few of biology's most famous double-edged swords—oxygen, insulin, and glucose—and glutamate is no different. Too much can cause *excitotoxicity*, harming nerve cells. Dysfunction of the complex mechanisms that govern glutamate release has been observed in Alzheimer's disease and is a destructive factor in ALS, a rapidly progressive neurological disease that attacks the neurons responsible for controlling voluntary movement. (One of the two major classes of drugs used to treat dementia reduces glutamate-related excitotoxicity in the brain, and the only FDA-approved drug to extend life in ALS is also a glutamate-modulating agent.)[3]

GABA

GABA inhibits the brain's overall level of excitation, and you're probably already familiar with how it feels to modulate it. Anti-anxiety drugs enhance the effect of GABA, as does alcohol, and both simultaneously inhibit glutamate. The problem is, these drugs are highly addictive and come with a host of consequences. The stimulating effects of caffeine on the other hand are owed to its ability to increase glutamate activity and inhibit GABA release. Anxiety, panic attacks, palpitations, and insomnia are all thought to manifest via dysfunction of the GABA system.

THE RISE OF THE "PSYCHOBIOTIC"

Scientists studying depression in mice have to be clever in how they determine what constitutes depression, and one of the many interesting ways to gauge the overall life satisfaction of a mouse is called the *forced swimming test*. Here's how it works: mice are dropped into a tank filled with water, where they immediately begin treading water until they find something to latch on to. Mice that are depressed tend to give up hope and allow themselves to sink sooner than happy mice, who tread water for much longer—this is interpreted as an increased motivation to live. As strange as this sounds, it's actually how some antidepressants are initially studied and tested.

In a unique twist on such an experiment, mice had their microbiomes populated with a certain type of probiotic called *Lactobacillus rhamnosus*, and then they were thrown into a tank. Compared to mice that didn't get the probiotic, those that did seemed more eager to stay afloat. They even showed a marked increase in anti-anxiety GABA receptors in certain parts of the brain. This effect was also absent in mice who *were* fed the probiotic but had their vagus nerves severed—the vagus nerve innervates the intestines and is connected directly to the brain. This suggests that the mechanism of action was direct microbial communication with the brain.[4]

If probiotics helped depressed mice, what are the odds they might help with other psychiatric symptoms? Mice born to mothers that ate the mouse equivalent of fast food (a deadly combination of fat and sugar) multiple times a day showed social behavior symptoms similar to autism. Upon inspection of their gut bacteria populations, these autistic mice had nine times less of another probiotic species, *Lactobacillus reuteri*, present. By restoring *L. reuteri* with a probiotic supplement, scientists were able to "correct" these social behavioral deficits, and the mice even showed increased production of the social hormone *oxytocin*, which acts like a neurotransmitter in the brain.

Interestingly, the amount of *L. reuteri* found in our systems has declined in tandem with the increases seen in rates of autism *and* consumption of fast food. In the 1960s, when the bacterium was discovered, *L. reuteri* was present in 30 to 40 percent of the population. Today it is found in 10 to 20 percent, a likely result of our diminished intake of fermented foods and fiber, our reliance on ultra-processed foods, and the rise in antibiotic use.[5] Considering that *L. reuteri* would normally be transmitted in breast milk, it's like the friend we didn't know we had until it was gone.

Optimizing Glutamate/GABA

One way of keeping this system functioning normally is to embody the glutamatergic/GABAergic balance in your own life, building in periods of deliberate excitation and inhibition. Intense exercise has been shown to promote balance, boosting both GABA and glutamate in the human brain.[6] This effect lasts well beyond the workout, as it correlates with higher resting levels of glutamate one week later. Major depression is categorized as having reduced levels of both, and exercise has been shown to improve depressive symptoms. Exercise has also been shown to help the brain more effectively metabolize glutamate, thus reducing its buildup.[7]

Meditation, yoga, and deep breathing exercises are excellent ways of achieving increased GABA.[8] Hypothermic conditioning, either by taking an ice bath or cold shower or by doing cryotherapy (where one enters a tank filled with freezing nitrogen gas, usually for around three minutes), is an excellent means of normalizing one's GABA/glutamatergic balance.[9] Though stressful and excitatory, thus stimulating the sympathetic "fight-or-flight" nervous response, people who participate in hypothermic conditioning will experience a significant drop in sympathetic activity and an increase in GABA after acclimation. (Cold exposure also has the benefit of boosting another neurotransmitter involved in learning and attention called *norepinephrine*, which I'll discuss momentarily.)

Avoiding consumption of added glutamate in processed foods is another strategy for keeping this vital balance of neurotransmitters. Monosodium glutamate (MSG), a flavor enhancer often used in Chinese cooking, is one common source, and aspartame, a noncaloric "diet" sweetener, becomes excitatory, transforming into glutamate precursors once inside the body.[10]

Acetylcholine: The Learning and Memory Neurotransmitter

Acetylcholine (pronounced *ah-see-till-KO-leen*) is a neurotransmitter that is part of the *cholinergic system*, which engages in many activities in the body but is mostly thought of for its role in REM sleep, learning, and memory.

Low levels of acetylcholine are associated with Alzheimer's disease, where acetylcholine-producing neurons become damaged. In fact, the second of the two major classes of drugs currently used to treat Alzheimer's disease and other dementias works to increase acetylcholine availability in the brain by preventing its enzymatic breakdown at the synapse* (I've already mentioned the first one, which modulates glutamate).

Optimizing Acetylcholine

One way to ensure optimal acetylcholine function is to avoid a broad class of very common "anticholinergic" drugs. Many of these drugs are widely used and available over the counter, used to treat everything from allergies to insomnia.

These drugs, as the word suggests, block the neurotransmitter acetylcholine, and continuous use can cause cognitive problems in as little as sixty days.[11] But even occasional use of a strong anticholinergic can cause acute toxicity. The symptoms are often remembered by med school students with the mnemonic "Blind as a bat (dilated pupils), red as a beet (flushing), hot as a hare (fever), dry as a bone (dry skin), mad as a hatter (confusion and short-term memory loss), bloated as a toad (urinary retention), and the heart runs alone (rapid heartbeat)."

Neurotransmitters are more than just the messages they contain—sometimes they are essential for keeping neurons healthy. Alarming research published in *JAMA Neurology* has shown that regular users of anticholinergic drugs had lower brain glucose metabolism and poorer cognitive abilities (including weaker short-term memory and executive function). Subjects even showed altered brain structures in MRI scans, displaying lower brain volume and larger ventricles (the cavities inside the brain). The anticholinergic drugs taken by these subjects included nighttime cold medicines, over-the-counter sleep aids, and muscle relaxants—all of which block acetylcholine.

You might be wondering if chronic use of these drugs can increase risk of dementia—and the answer is yes. In a study of 3,500 older

---

* These drugs, called *cholinesterase inhibitors*, are not known for being particularly efficacious, in part because low acetylcholine is the result of underlying dysfunction, not the cause of it. These drugs do nothing to treat that dysfunction, and therefore do not alter disease progression.

COMMON ANTICHOLINERGIC DRUGS TO AVOID

Drug	Use	Impact
Dimenhydrinate	Motion sickness	Strong anticholinergic
Diphenhydramine	Antihistamine/sleep aid	Strong anticholinergic
Doxylamine	Sleep aid	Strong anticholinergic
Paroxetine	Antidepressant	Strong anticholinergic
Quetiapine	Antidepressant	Strong anticholinergic
Oxybutynin	Overactive bladder	Strong anticholinergic
Cyclobenzaprine	Muscle relaxant	Moderate anticholinergic
Alprazolam	Anti-anxiety	Possible anticholinergic
Aripiprazole	Antidepressant	Possible anticholinergic
Cetirizine	Antihistamine	Possible anticholinergic
Loratadine	Antihistamine	Possible anticholinergic
Ranitidine	Anti-heartburn	Possible anticholinergic

adults, University of Washington researchers found that people who used these drugs were more likely to have developed dementia than those who didn't use them.[12] In fact, the more regular the usage, the greater the dementia risk. Taking an anticholinergic for the equivalent of three years or more was associated with a 54 percent higher dementia risk than taking the same dose for three months or less. If you take any of these drugs regularly, it is critical that you have a conversation with your physician about the possibility that they may be impairing your cognitive function and ultimately putting you at higher risk for dementia. If you are a carrier of the *ApoE4* allele (described in chapter 6; carriers make up 25 percent of the population) or have a strong family history of dementia, you and your doctor should definitely look for safer alternatives.

Diet also factors into an optimal cholinergic system. Choline is the major dietary precursor to acetylcholine, and changes in plasma levels of choline lead to changes in brain levels of the precursors for this neurotransmitter.[13] Choline is also a key component of cell membranes, which is where the body stockpiles it for later use. It's found in abundant quantities in seafood and poultry, but eggs may be the best source, with one large egg containing about 125 milligrams of choline in its yolk. Unfortunately, average choline intake in the United States is far below the adequate level, set at 550 milligrams per day for men and 425 milligrams per day for women (higher for those who are pregnant or breastfeeding) by the National Academy of Medicine.[14] Ten percent or fewer had choline intakes at or above these levels.[15]

Top Choline Foods

- Eggs (eat your yolks!)
- Beef liver
- Shrimp
- Scallops
- Beef
- Chicken
- Fish
- Brussels sprouts
- Broccoli
- Spinach

(Vegans and vegetarians: you'd need to eat two full cups of broccoli or Brussels sprouts to get the same amount of choline as found in one single egg yolk.)

Serotonin: The Mood Neurotransmitter

Growing up in New York City, I would always feel my spirits drop come the fall. The impending winter months of long, dark days with little sun exposure gave rise to a kind of depression known as *seasonal affective disorder*, or SAD. Also known as *winter depression*, SAD affects an estimated ten million people in the United States—and while the majority of those affected are women, everybody is at risk.

When I was seventeen, I learned that the skin made vitamin D from exposure to the sun, and I realized that my lack of sun exposure during those dark months was likely compromising my body's vitamin D production. I had a hunch that my mood, the limited sun I was receiving, and the reduced vitamin D that I was synthesizing might all somehow be related. So I self-prescribed a vitamin D supplement to see if that would improve my mood. Lo and behold, I felt better.

Was this a placebo effect? One can never be sure—a double-blind trial this was not. However, nearly two decades since my experiment, scientists have discovered a mechanism that might very well explain the improvement I felt. It turns out that healthy serotonin levels may actually rely on vitamin D, as vitamin D helps to create serotonin from its precursor, the amino acid *tryptophan*. This is an important insight, particularly in light of research estimating a vitamin D deficiency in three-fourths of the US population.

Serotonin is well known for its mood- and sleep-boosting abilities, and it makes up the basis of what is called the *serotonergic system*. You may be familiar with the class of antidepressant drugs on the market called *selective serotonin reuptake inhibitors*, or SSRIs. These drugs promise to boost serotonin's availability at the synapse by preventing its reuptake into the presynaptic cell.

Again, prescription drugs aren't the only compounds that tinker with this neurotransmitter. The drug MDMA is known for its mood-altering influence, attributed to its effect on the serotonergic system. Initially studied for its potential to treat post-traumatic stress and other treatment-resistant mental disorders, MDMA is like dynamite to the dam governing normal serotonin release. But the act of releasing huge amounts of serotonin overwhelms the recycling machinery and causes oxidation of the surrounding neurons, literally burning them away—perhaps this is why chronic, long-term MDMA use has been linked with memory problems and brain damage. (A recurring theme in this book is that every biologic action has an equal and opposite reaction—there's no such thing as a biological free lunch!)

Another compound, psilocybin, the psychoactive chemical in "magic" mushrooms, prevents serotonin reuptake and also mimics serotonin, activating its receptors. This is unlike MDMA, which floods the synapses with your own serotonin. For this reason, psilocybin may have fewer negative long-term effects. In groundbreaking research performed by both New York University and Johns Hopkins University, psilocybin was shown to alleviate anxiety and increase feelings of life satisfaction in patients with life-threatening cancer for six months after just a single dose.[16] The cognitive-enhancing potential of low doses of psilocybin, called *micro-dosing*, is currently being studied.

Serotonin isn't just for good vibes. It's also heavily involved in executive function. We know this because scientists have devised a clever way of temporarily reducing levels of serotonin in people, and the results aren't pretty. As I've mentioned, serotonin is synthesized in the brain from the essential amino acid *tryptophan*. The tryptophan we consume in protein makes its way up to the brain, where it then has to be carried across the blood-brain barrier by transporters. Other amino acids vie for these same transporters, including branched-chain amino acids, which are important for brain function and muscle growth, among other things. When given as a standalone supplement, these amino acids can outcompete tryptophan, blocking its entry into the brain.[17] (One of the ways exercise actually boosts your mood is by causing your muscles to "suck up" circulating branched-chain amino acids, thus allowing tryptophan *easier* entry into the brain—more on this in a moment.)

So, what happened when scientists gave these amino acids to subjects? Serotonin levels temporarily plummeted. This was accompanied by a wide range of behavioral changes including increased aggression, impaired learning and memory, poorer impulse control, a reduced ability to resist short-term gratification, impaired long-term planning, and reduced altruism.[18] It's easy to see how these traits might reinforce feelings of depression, and even contribute to violent tendencies. As a fascinating aside, bright light exposure during such a treatment seems to attenuate some of these effects, suggesting yet another method of ensuring healthy serotonin expression: daily sunlight exposure.[19]

Optimizing Serotonin

By this point, you are starting to understand how to keep inflammation in your body to a minimum: avoiding sugars, grains, and oxidized oils, while consuming ample plant nutrients and fiber (we'll cover more in the following chapters before we put it all together in the Plan). If you've already begun integrating these ideas into your lifestyle, you're on your way to optimizing your serotonin expression. This is because inflammation may block serotonin from being released by your neurons, as was shown in research from Children's Hospital Oakland Research Institute (CHORI).[20] This perhaps explains why depression caused by chronic inflammation is resistant to traditional therapy methods, but can be treated by reducing inflammation in the body. The CHORI researchers found that the anti-inflammatory omega-3 fatty acid EPA facilitated normal serotonin release while DHA, which supports membrane fluidity (discussed in chapter 2), promoted healthy uptake into the postsynaptic cell.

As SSRI sales continue to skyrocket, this type of research is increasingly important, providing strong evidence that "low serotonin" may be the result of an underlying problem for many people, rather than the cause of depression itself. This kind of insight is sorely needed, especially considering that one in ten Americans now takes an antidepressant medication; among women in their forties and fifties, the figure is one in four.[21] And are they effective? A recent *JAMA* meta-analysis concluded:

> The magnitude of benefit of antidepressant medication compared with placebo increases with severity of depression

> symptoms and may be minimal or nonexistent, on average, in patients with mild or moderate symptoms. For patients with very severe depression, the benefit of medications over placebo is substantial.[22]

In other words, for a lot of people, antidepressants are no better than a placebo, with the exception being the most severe cases of depression (and even then, nonpharmacologic treatments have shown impressive successes, like the recently published SMILES trial—see sidebar for more—as well as experiments involving anti-inflammatory compounds such as curcumin, a component of the spice turmeric).[23]

CAN DIET *REALLY* TREAT DEPRESSION? THE SMILES TRIAL

The link between depression and poor diet is well established, and being depressed can certainly lead us to unhealthy eating. But do poor diets make us depressed, and if we improve our diets, will our mental health improve? We now have an answer, thanks to the SMILES trial, published in 2017 and led by Dr. Felice Jacka, the director of Deakin University's Food and Mood Centre in Australia.

Using a modified Mediterranean diet focusing on fresh vegetables, fruits, raw unsalted nuts, eggs, olive oil, fish, and grass-fed beef, Jacka and team saw patients with major depression improve their scores by an average of about eleven points on a sixty-point depression scale. By the end of the trial, 32 percent of patients had scores so low that they no longer met the criteria for depression! Meanwhile, people in the group with no dietary modification improved by only about four points, and only 8 percent achieved remission.

This data adds serious strength to the argument that we can eat our way to a better mood, and the Genius Plan has been calibrated to incorporate these findings. To watch an hour-long, in-depth interview that I conducted with Dr. Jacka, visit http://maxl.ug/felicejackainterview.

Aside from sunlight, vitamin D, and omega-3 fatty acids DHA and EPA, what other magic supplement might you take to boost serotonin? Given the body's incredible ability to synthesize neurotransmitters from the most basic of building blocks, the most potent means we know of boosting serotonin in the brain is to simply *move.* As I mentioned earlier, exercise boosts plasma tryptophan (remember, this is serotonin's precursor) and decreases levels of the branched-chain amino acids, which, though important, compete with tryptophan for entry into the brain. This substantial boost in tryptophan's availability to the brain persists even after the workout is finished.[24] In another powerful and illuminating head-to-head study, no SSRI was as effective as exercise, three times per week, in combating depression. That's *game*, *set*, and *match*!

There is another means of boosting serotonin in the brain that you may already be familiar with, and that is with carbohydrates and sugar. This temporary mood boost is behind one of the more addictive qualities of carbohydrates. And then when carbohydrate levels curtail in between meals, serotonin drops, sending us reaching for something starchy or sugary—demonstrating why carbohydrate consumption is not a sound strategy for increasing serotonin.

In pivotal psychological studies, feeding sugar to subjects has been shown to temporarily improve willpower and executive function, but it's hard to tease out whether the sugar actually boosts function or simply treats withdrawal from the lack of it. It would help to see these studies replicated on fat-adapted subjects. Regardless, though, short-circuiting the reward systems of the brain with external stimuli, whether with sugar, drugs, sex, or chronic bouts of extended and high-intensity cardio, rarely leads anywhere positive in the long term. But sugar in particular keeps you on an insulin roller coaster throughout the day, can cause you to gain weight, and can lead to metabolic dysfunction, reinforcing many of the inflammatory mechanisms making you depressed to begin with!

SEROTONIN AND THE GUT

According to an oft-cited statistic, 90 percent of the body's supply of serotonin is found in the gut, not the brain. This is true, as gut epithelial cells create serotonin, facilitating digestion. Does the key to happiness, then, lie in the gut? You bet—but the reason why may surprise you. Gut-derived serotonin does not cross the blood-brain barrier. But at the same time, what happens in the gut can influence brain serotonin activity via its ability to modulate inflammation.

In chapter 7, we discussed gut health and the necessity for maintaining gut barrier integrity by consuming soluble fiber–containing vegetables that help to "seal up" the gut's pores. Lipopolysaccharide (LPS), a normal constituent of a healthy bowel, becomes a potent instigator of inflammation when seeped through a "leaky" gut. Aside from sending the immune system into an inflamed state of defense, LPS is directly toxic to both the serotonin and dopamine systems. In fact, LPS is often used in the lab setting, injected into mice to induce depressive behavior and neurodegeneration. Revisit chapter 7 for ways to protect and enhance your gut wall integrity.

Dopamine: The Reward and Reinforcement Neurotransmitter

Like serotonin, dopamine is considered a "feel-good" neurotransmitter. It is most famously associated with motivation and reward, and it gets released when we do things like have sex, listen to our favorite music, eat, or watch our favorite sports team succeed. It also spikes when a new work opportunity or promotion presents itself, when we spot someone from across the bar that we find attractive, or when we get a notification of "likes" on social media posts. When goals are set and met, our dopamine system lights up, helping to motivate us to do things that evolution has deemed good for us and the species. But this system, like many others, can become dysfunctional in the modern world.

Because of dopamine's role in motivation, it is heavily involved in aspects of executive function, where it mediates motor control, arousal, and reinforcement. Its presence is reduced in addicts, who attempt to normalize their levels with substances or actions. This is one of the reasons "uppers" are so addictive, as they commonly increase levels of dopamine in the brain, via a variety of mechanisms. Cocaine, for example, inhibits dopamine reuptake, resulting in increased dopamine concentrations in the space between neurons. Methamphetamine, on the other hand, causes a flood of dopamine from the presynaptic neuron, while also preventing its reuptake. "Crystal meth," one form of methamphetamine, is highly neurotoxic, killing off natural dopamine-producing cells, which compounds its highly addictive nature (and keeps Walter White in business).

In Parkinson's disease, the dopamine-producing cells of a specific part of the brain called the *substantia nigra* become damaged, so patients can take dopamine-boosting drugs that alleviate symptoms for a time. Eventually, these medications lose effectiveness in part because the artificial flood of neurotransmitters causes the downregulation (or lessening) of dopamine receptors throughout the brain.[25]

This is actually a self-regulating mechanism that all neurons have to reduce or increase the cell's sensitivity to neurotransmitters, but it is especially risky with dopamine. One peculiar side effect of dopamine-enhancing therapies in Parkinson's disease is a potential increase in "risky behavior," including pathological gambling, compulsive sexual behavior, and excessive shopping.

Dopaminergic activity is also reduced in ADHD, where there are fewer dopamine receptors in the postsynaptic cell. This means that more dopamine is required to sustain attention and focus. But is this a disorder, or simply a brain that is hardwired for novelty seeking?

ARE YOU A WORRIER OR WARRIOR?

Certain genes modulate neurotransmitter function and thus play a role in key aspects of personality. The *COMT* gene is one of the more well-studied genes of the bunch. It is responsible for producing catechol-O-methyltransferase (COMT), an enzyme that breaks down dopamine in the prefrontal cortex, the part of the brain responsible for higher cognitive and executive function.

Each of us inherits either two As, two Gs, or an A and a G. These letters represent variations called *alleles*. Having two A copies results in a three- to fourfold decrease in COMT enzyme activity compared to the G allele, and having a combination of the two splits the difference. Depending on what variation you have, dopamine either gets broken down more quickly or more slowly at the synapse. Therefore, if you have two As, then you will have more dopamine hanging around in the prefrontal cortex under everyday circumstances (because dopamine is broken down more slowly), while carrying two Gs results in the lowest amount of dopamine (because dopamine is broken down more quickly). Those with AG fall somewhere in the middle.

The A allele is thought to be the "worrier" allele, and people who carry two copies tend to be more neurotic and less extroverted. When worriers experience a spike in dopamine, they really feel it—this is why AAs tend to feel "higher highs" and get the sense that they are getting more out of life. Though high dopamine may seem like a good thing, too much postsynaptic stimulation can cause cognitive performance to suffer. Worriers perform poorly under stressful conditions for this reason, but display better cognitive performance under normal conditions. AAs also tend to experience lower lows and display less emotional resilience, being more prone to anxiety and depression. On the other hand, they are thought to be more creative.

The G allele is thought to be the "warrior" allele, and people who carry two copies tend to be less neurotic and more extroverted. Warriors handle stressful situations much better than worriers, retaining peak cognitive prowess during times of stress and uncertainty. They also display greater emotional resilience and have better working memory. They can be more cooperative, helpful, and empathic. On the other hand, these noble warriors tend to feel as though they're getting less out of life.

As you can see, both allele patterns embody personality characteristics necessary for the success of any tribe, and carriers of one A and one G (called *heterozygosity*) have personality traits somewhere in between warrior and worrier, embodying the best—and worst—of both worlds. To learn your status, sign up for a consumer gene testing service that provides access to raw data and search for SNP rs4680. Just remember: whether you are a warrior or worrier, the terms are mere generalizations. Everyone is unique. Case in point: yours truly is a warrior with a creative streak!

A recent *New York Times* op-ed suggested that the preference for novelty seeking in the ADHD brain may be a trait that, until relatively recently, served a distinct evolutionary advantage for our species, having evolved over millions of years as nomadic hunter-gatherers.[26] It makes perfect sense: a successful hunter-gatherer would need to be motivated to seek out new foraging opportunities, and to be rewarded by her brain once she found them. In today's world of assembly-line education and highly specialized career choices, those with ADHD may be silently suffering from the "tranquility of repetition," to borrow a line from one of my favorite films,[†] often ending up on drugs like Adderall (dextroamphetamine) and Ritalin (methylphenidate). These drugs, like cocaine, are dopamine reuptake inhibitors.

Clinical professor of psychiatry at Weill Cornell Medicine and the op-ed's author Richard Friedman wrote of one of his successful patients: "[He] 'treated' his ADHD simply by changing the conditions of his work environment from one that was highly routine to one that was varied and unpredictable." This may explain why a disproportionate number of those with ADHD and learning disabilities gravitate toward more entrepreneurial careers.[27]

FIGHTING "HEDONIC ADAPTATION" (AKA THE HUMAN CONDITION)

A common problem with dopamine is that we can become tolerant to its effects when stimulated. This is clearly observed by the phenomenon of hedonic adaptation. Think of a former life goal that you've already achieved. Perhaps it was buying a car you'd always wanted, or getting that promotion, or moving into a new house. Certainly, these are incredible and exciting life milestones, but as sure as you are human, your level of happiness regressed to baseline after the initial buzz wore off. This "tolerance" to dopamine, especially when achieved by any short-circuiting of the stimulus/reward pathways in the brain, can result in "anhedonia," or a pathological *inability* to sense or experience pleasure toward things that

[†] That's *V for Vendetta*—a flawless movie in my opinion. *Remember, remember* . . .

we previously found enjoyable. But there is a solution: *absence makes the dopamine receptor grow fonder.*

Buddhist monks have known for centuries that abstention provides a means of stepping off the hedonic treadmill. Any prolonged reduction of dopamine release will cause an upregulation of receptors, thus increasing sensitivity to dopamine. While asceticism may not work for everyone, taking a deliberate "time-out" from otherwise dopamine-reinforcing habits—technology usage, for example—can be an incredibly effective way of boosting motivation, reestablishing healthy relationships, and enhancing overall happiness.

Not ready to totally unplug? Try this simple happiness hack for a week: make a rule of no computers, e-mails, or texts for one hour after waking up, and the same one hour before bedtime. As your system resets, you may be inclined to stick with it.

Optimizing Dopamine

Dopamine is made in the brain from the amino acid *tyrosine*, and as with other neurotransmitters, the building blocks are usually readily available unless a person is protein deficient. In this sense, a healthy dopaminergic system may be more a function of our choices and actions than any nutrient deficiency. Consuming foods that have been processed to become hyper-palatable, engaging in risky activities, and taking substances that hijack and short-circuit the brain's reward system can create unhealthy and self-destructive addictions. Sugar and rapidly digested carbohydrates like wheat are massive dopamine stimulators, evolved to increase fat storage at a time of seasonal sugar availability. The addictive qualities of sugar are so strong, they are often compared to some of the illicit drugs I've mentioned above.[28] Even the feedback loops created by social media—certainly a positive force in many regards—can cause the dopamine system to become dysregulated and promote addiction.

Conversely, setting both short- and long-term goals for yourself is a nice "hack" that allows anticipation (an important aspect of sustained happiness) and reward. Try picking up a novel exercise routine, learning to play a new instrument, getting out of your social comfort zone, falling in love, or beginning a side entrepreneurial project. These are all ways of healthily boosting dopamine.

Norepinephrine: The Focus Neurotransmitter

Even though dopamine and serotonin are arguably the more well-known neurotransmitters, norepinephrine is equally noteworthy. Norepinephrine plays a very important role in focus and attention and is expressed in the brain whenever focus is needed, particularly in times of stress where it can enhance long-term memory formation. Can you remember where you were the moment you heard of the 9/11 attacks on the World Trade Center? I'll bet that day is burned into your memory in crystal-clear, stunning detail. This is owed to none other than norepinephrine.

The main hub of norepinephrine is a small region in the brain known as the *locus coeruleus*. Any stressful stimuli lead to an increase in norepinephrine, from a terrorist attack to a major fight with a significant other to simply not eating for twenty-plus hours. Evolutionarily speaking, this is an important adaptive function. For much of our time on the planet, stressful stimuli required our immediate attention, and detailed, long-lasting memories needed to be formed to avoid such an event in the future (provided that we lived through that initial encounter). This is called *long-term potentiation*, and it plays an important role in fear conditioning. Because norepinephrine has such a powerful effect, undoing learned fear can be an intensive process—ask anybody suffering from post-traumatic stress disorder, or PTSD.

Milder forms of stress can also activate many of the same pathways. The "stress" of learning a new instrument, solving a crossword puzzle, or experiencing *novelty*—exploring a new town, or going on a walk with changing scenery, for example—all have been shown to increase norepinephrine in the brain. This can be very beneficial, as norepinephrine helps to make the connections between neurons stronger.

DOCTOR'S NOTE: WALK AND TALK FOR ENHANCED MEMORY

When patients and I get to know each other, I often conduct my visits while walking with the patient in New York City's Central Park. The movement and constantly changing scenery helps the patient remember my advice, and helps me remember the encounter!

Optimizing Norepinephrine

Norepinephrine can sometimes work against us. Unlike in our distant past, stressful stimuli today don't always require our immediate focus and attention. And yet, the physiological mechanisms remain to guide our attention when a threat is perceived, real or otherwise. The media often exploits this fact—something I know well, having worked in TV. "If it bleeds, it leads" is typically the mandate in television news, where the most stress-inducing stories are teased at the top of the hour. This approach activates networks in our brains that ensure that attention is paid to such news as if our survival depends on it. Clearly, this is often not the case. In fact, avoiding daily news is one strategy to enhance your focus and cognition that will pay dividends, as chronic norepinephrine release can harm your cognitive function just as much as acute release can boost it.

How can we mine norepinephrine for greater productivity? Exercise is one of the most effective ways of boosting norepinephrine, and the "side effect" of that may mean enhanced learning and memory. This was demonstrated recently when college-age adults who exercised on a stationary bike while learning a new language were able to retain and understand what they learned better than controls who were sedentary during lessons.[29] For the millions diagnosed with ADHD, this strategy can serve as a natural cognitive enhancer, as norepinephrine (and dopamine) reuptake inhibitors are frequently prescribed to treat ADHD.

While the vast majority of medical school curricula are absent any focus on exercise for their doctors-to-be, research has borne out that exercise may be the best form of medicine for the ADHD brain:[30] in a slew of trials, kids who took part in a regular physical activity program showed enhanced cognitive performance and executive function and improved math and reading test scores and demonstrated an overall reduction of ADHD symptoms. Perhaps we should keep this in mind when physical education appears on the chopping block at the next school budget meeting.

NERD ALERT

As Dr. Paul and I work on this chapter, we are taking turns writing and doing kettlebell swings. We both often used this "hack" in school when studying for a grueling exam. This not only would help break up the marathon of monotony by doing push-ups, squats, or dips between two chairs in the library, but would also serve as a way to increase blood flow, enhance mental acuity, and intimidate our peers.

Interestingly, extreme temperatures are another type of physical stressor that can induce effects similar to exercise. One study found that when men stayed in a sauna that was heated to 80°C (176°F) until subjective exhaustion, their levels of norepinephrine increased threefold.[31] (A study in women saw a similar but smaller boost.)[32] Cold water immersion is another massive neurological modulator and has been used by various cultures for centuries as a health-boosting tool. Anyone feeling mentally fatigued need only take a cold shower or have a dip in an ice bath to reap the mental benefits—the surge of norepinephrine from a cold shock, reported to increase over fivefold in humans, can make the brain feel like it's coming back "online."[33] It's perhaps not a coincidence that a major Russian holiday that involves plunging into freezing lakes through holes cut in the ice is called Epiphany!

Aside from its well-documented role in focus, attention, and memory formation, some other very interesting aspects of norepinephrine have emerged from the literature. Increasing norepinephrine in animals boosted their resilience to stress, enhancing their capacity to rebound from traumatic events.[34] Norepinephrine also has an anti-inflammatory effect in the brain, and stimulating the neurotransmitter may strengthen a brain region that tends to be implicated early in the development of Alzheimer's disease.[35] A team of researchers out of the University of Southern California have called the locus coeruleus, the main hub of norepinephrine release, the "ground zero" of Alzheimer's disease. In Alzheimer's, up to 70 percent of norepinephrine-producing cells are lost, and the decline of norepinephrine tightly correlates with the progression and extent of cognitive impairment.[36] The researchers demonstrated in rodents that causing norepinephrine to be released helped protect neurons from inflammation and excessive stimulation (both key players in Alzheimer's disease).[37]

Optimizing the System

Now that you know how to optimize each neurotransmitter, let's go over some practical steps you can take to ensure that your whole brain is functioning at its peak at any given time.

Protect Your Synapses

The point where neurons meet is called a *synapse*. Each neuron may hold connections with up to ten thousand other neurons, passing signals to one another via as many as one thousand trillion synaptic connections.[38] (This is modest in comparison to a child's brain: a three-year-old may have one quadrillion synapses!) Keeping these connection points healthy by minimizing excessive oxidation is key to the optimization of your many cognitive processes. In fact, synaptic dysfunction is an early marker of Alzheimer's disease, and the typical decline of certain aspects of cognitive performance with age may also be a result of synaptic dysfunction.[39] Protect your synapses from oxidative stress by doing the following:

- Consume DHA fat from fatty fish, or consider supplementing with high-quality fish oil.
- Avoid the consumption of polyunsaturated oils (revisit chapter 2) and increase consumption of extra-virgin olive oil.
- Consume ample fat-soluble antioxidants like vitamin E (found in avocados, almonds, and grass-fed beef), carotenoids like lutein and zeaxanthin (found in kale, avocados, and pistachio nuts), and astaxanthin (found in krill oil).

Express Your Inner "Wonder Junkie"

I love the term "wonder junkie," which I discovered upon reading Carl Sagan's novel *Contact*. It's a term he uses to describe the central character, Ellie Arroway, who dedicates her life to exploring the unknown. New experiences stimulate *synaptogenesis*, or the creation of new synapses, while the loss of synaptic connections coincides with memory loss.[40] Get out of your comfort zone and explore new and unfamiliar territories. Stagnation is death—particularly where your brain cells are concerned.

Avoid Toxic Chemicals

One of the potential factors affecting neurotransmitter function is the consumption of toxic pesticide residues, which are nearly ubiquitous in the modern food supply. Pesticides work by causing swift and irreparable damage to the nervous systems of insects, particularly affecting the cholinergic system (important for learning and memory). Although it would take massive, concentrated exposure to have this effect in humans, it is not implausible that lifetime exposure to low levels through contaminated food could alter the function of our neurotransmitters—a scientific unknown at this point.

There is also strong evidence linking pesticides and herbicides to Parkinson's disease, the most common neurodegenerative disease after Alzheimer's, where the dopamine-producing cells of the *substantia nigra* die. In human studies, people who have been exposed to large amounts of these chemicals display dramatically increased risk for Parkinson's, with exposure to some fungicides associated with twofold risk![41] Parkinson's disease etiology—what causes it to develop—is not yet understood, but toxic exposure is a leading suspect.[42]

Pesticides can also cause harm in developing fetuses via a similar lever. Studies in lab animals using model compounds suggest that many pesticides currently in use can cause adverse effects on brain development. Curiously, current safety requirements for pesticide use do not include testing for developmental neurotoxicity.

To be clear, the verdict on how this all affects humans is not yet certain, and it could be years before science arrives at concrete conclusions. Given the astonishing degree of commerce tied to these issues, expecting firm answers any time soon may be wishful thinking, as the research is sure to be mired in the eternal conflict between corporate interest and scientific inquiry. But by opting for organic produce we may be able to at least stack the odds in our own favor as we await greater clarity.

Engage in Brief Periods of Fasting

New research from the Buck Institute for Research on Aging suggests that fasting is able to "tune" synaptic activity, giving these high-activity connection points a rest as an energy conservation measure. When researchers observed the neurons of fruit fly larvae in a fasted state, the amount of neurotransmitters released decreased dramatically, which essentially cleaned up the synaptic gaps. This is positive because excess neurotransmitters left to linger in the synaptic gaps can generate damage-causing free radicals. Fasting may therefore help limit unwanted oxidative damage in the brain (alongside its ability to reduce the brain's dependency on glucose).[43] Chapter 10 revisits practical and safe fasting protocols.

Avoid Excessive Stimulation

Multiple systems are involved in mediating sensory input. When our senses become overloaded, this can substantially diminish our executive function. This is perfectly illustrated by what happens when we watch a movie. The sights and sounds completely envelop us and we feel immersed in the universe of the film. This occurs because intense sensorimotor processing inhibits parts of our brain responsible for self-awareness.[44] During a movie, this is exactly what we want—cinema is meant to be, after all, a dream shared by audience and director. But in day-to-day life, unintentional overload can occur at the expense of our executive function. The modern world can be excessively excitatory—music, electronic billboards, the light from a smartphone screen, the flickering of a TV screen, or simply the sound of a train entering the station. These are all factors that, when combined, can overload our prefrontal cortexes and deplete our neurotransmitter stores.

Here are some ways to reduce excessive stimulation:

- **When requiring focus, such as when working or studying, ensure that any music you choose to listen to is instrumental only.** Lyrics engage the brain's language center, which can compromise your ability to use language in other tasks simultaneously.
- **Turn the volume down on your devices (TV, smartphone, etc.).** Keep the volume as low as possible while still being able to enjoy your content.
- **Turn down the brightness on your screens.** Many people keep their smartphone screen brightness up to the maximum setting. Set yours to adjust automatically to ambient light, and keep it on minimum brightness at night.
- **Use warmer-colored bulbs in your home.** Bulbs that give off a more "orange" glow contain less of the blue wavelengths of light, which can overstimulate the brain in evening hours.
- **Eliminate overhead lighting, especially at night.** Overhead lighting signals to the brain that the sun is out. Eye-level lighting in the evening hours (from lamps, for example) is much more soothing to a brain that is trying to wind down. We had fires to light our evenings for four hundred thousand years (some estimate even longer) but overhead light bulbs for less than two hundred.[45]
- **Meditate.** I recommend being properly trained in meditation. Whatever style you choose, research shows it to be a very smart investment. Take the time to do it right, and you'll have it with you for the rest of your life! I've written an entire beginner's guide to meditation (along with recommendations for good online courses) on my website at http://maxl.ug/meditation.

Next up in our journey to discover the optimal brain-building lifestyle, we are going to look at hormones—which, along with neurotransmitters, guide our moment-to-moment decisions. They are involved in everything from mood to metabolism, and understanding them will be the final piece in understanding the role of nutrition in attaining your ultimate cognitive birthright.

FIELD NOTES

YIN AND YANG (GLUTAMATE AND GABA)

- Acknowledge your biological excitation mode and inhibition mode, and understand that you need both regularly: exercise and recovery, adventure and relaxation.

ACETYLCHOLINE

- Avoid toxic anticholinergic drugs.
- Ensure adequate dietary choline intake.

SEROTONIN

- Maintain adequate omega-3 intake (see chapter 2 for a refresher).
- Get a blood test to ensure optimal vitamin D levels. Most often you will have to specifically request that your doctor test for this, and it is an inexpensive test. Though there is no consensus, the latest research suggests that having D levels in the range of 40 to 60 ng/ml is optimal (see chapter 12 for further explanation).
- Exercise often, which sends tryptophan straight to the brain, in an effect that persists even after exercise.

- Without looking directly into the sun, ensure daily bright sunlight exposure. Even on a cloudy day, the light outside is brighter than anything you could achieve indoors and is sufficient to improve mood.
- Follow the gut health plan set forth in chapter 7.

DOPAMINE

- Begin a novel exercise routine.
- Learn to play a new instrument.
- Get out of your social comfort zone.
- Begin a side entrepreneurial project.
- Start a new blog, newsletter, or Meetup group.
- Break from the "tranquility of repetition." Take alternate routes to work and travel more frequently.

NOREPINEPHRINE

- Halt chronic consumption of news, which causes, more often than not, an unnecessary spike in norepinephrine.
- When you require long periods of focus, engage in short, frequent bouts of physical activity.

GENIUS FOOD #8

BROCCOLI

Our moms were right. Broccoli and other cruciferous vegetables (including Brussels sprouts, cabbage, radishes, arugula, bok choy, and kale) are very beneficial to our health, in part because they are dietary sources of a compound called *sulforaphane*. This powerful chemical is created when two other compounds, held in separate compartments of these plant's cells, unite as a result of chewing.

Sulforaphane is currently being studied for its impact on a variety of conditions, and it has already shown tremendous promise in treating or preventing cancer, autism, autoimmunity, brain inflammation, gut inflammation, and obesity. One fascinating study showed that mice fed sulforaphane along with an obesity-promoting diet gained 15 percent less weight and had 20 percent less visceral fat compared to mice that weren't fed sulforaphane with their fat-inducing diets. Sulforaphane is not a vitamin or an essential nutrient. Instead, sulforaphane is a powerful genetic modulator known for its activation of an antioxidant pathway called *Nrf2*. *Nrf2* is the body's master switch for creating powerful chemicals that mop up oxidative stress. While other beneficial compounds such as plant polyphenols also stimulate this pathway, sulforaphane is the most potent of the known *Nrf2* activators. That begs the question: what is the top known source of sulforaphane?

Youth has its perks, especially if you're broccoli. Young broccoli sprouts yield anywhere between twenty to one hundred times the sulforaphane-producing compounds as adult broccoli (if we're speaking strictly in terms of micronutrient content, adult broccoli is still more nutritious than broccoli sprouts). One pound of sprouts therefore equals one hundred pounds of adult broccoli in terms of its sulforaphane-producing capacity.

How to use: Add cruciferous vegetables to your diet and consume them raw and cooked. Just note that one of the two compounds that creates sulforaphane (an enzyme called *myrosinase*) is destroyed by high-heat cooking. Thus, cooked broccoli and other crucifers lose their ability to create sulforaphane upon chewing. However, you can add myrosinase after the fact. Mustard powder is particularly rich in this compound, and if you sprinkle some onto your veggies after they've been cooked, voilà—the ability to create sulforaphane is regained![1]

Pro tip: Growing your own broccoli sprouts is incredibly cost-effective and easy, even for those of us without a green thumb. Visit http://maxl.ug/broccolisprouts for my step-by-step guide on how to grow broccoli sprouts in just three days using the easiest method I've found. Blend them into smoothies, use to top grass-fed beef or turkey burgers, or add generously to salads.

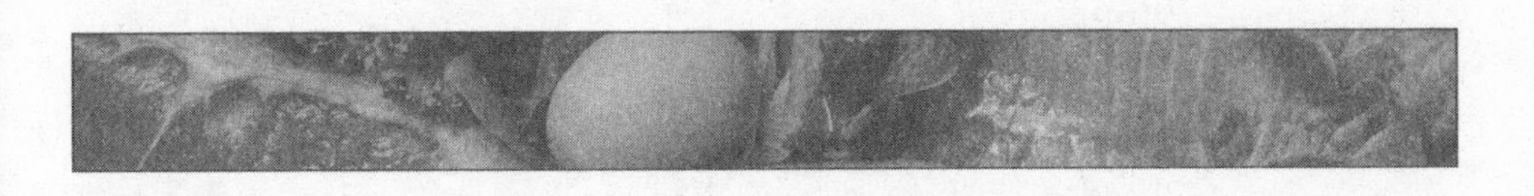

PART 3

Putting Yourself in the Driver's Seat

CHAPTER 9

SACRED SLEEP (AND THE HORMONAL HELPERS)

Do but consider what an excellent thing sleep is . . . that golden chain that ties health and our bodies together. Who complains of want? of wounds? of cares? of great men's oppressions? of captivity? whilst he sleepeth? Beggars in their beds take as much pleasure as kings: can we therefore surfeit on this delicate Ambrosia?

—THOMAS DEKKER, PLAYWRIGHT

Want a "biohack"? Here's one: go to sleep.

I know, I know. *Easier said than done, Max. I run a business! I'm in grad school! I have two and a half kids! I'm not caught up on* Game of Thrones*!* I get it. We all have careers, obligations to our friends and families, creative endeavors, shows we want to watch, and of course the upkeep of our precious Instagram-Facebook-Twitter-Snapchat-Tinder accounts. But as you're about to see, sleep controls the tide on all the ships in your harbor, and a good night's sleep raises them all. It solidifies our memories, boosts our creativity, increases our willpower, and regulates our appetite. It resets our hormones, gives our neurons a cleansing bath, and ensures "all systems go" in the various regions of our infinitely complex brains. It's no wonder we intuitively know to "sleep on it" before making an important decision.

A sleepless brain, on the other hand, is like marooning your ships on the beach at low tide. New research even pegs sleep loss as a toxin to your energy-creating mitochondria, putting it in the same category as processed oils and sugar.[1] In one study published in the journal *Sleep*, a single night of sleep deprivation in healthy human volunteers led to a 20 percent increase in two markers of neuronal injury, suggesting that even one instance of acute sleep deprivation may cause injury to your precious brain cells.[2]

This is alarming news in light of the fact that half of adults between twenty-five and fifty-five say they sleep fewer than seven hours on weeknights.[3] And more than 50 percent of millennials have been kept awake at least one night over the past month due to stress—a recent finding by the American Psychological Association.[4]

A SLEEPLESS BRAIN IS *PRIMAL*—AND NOT IN A GOOD WAY

Ever had the sensation of "losing yourself" in a great movie, book, or video game? What about a workout session, sex, or playing your favorite instrument? We owe this incredible, life-affirming sensation of complete immersion to a relative disengagement of the prefrontal cortex. Located at the very front of your brain, just behind your forehead, the prefrontal cortex is thought to be responsible for planning, decision making, expression of one's personality, and self-awareness itself. Save for when we send it on vacation with the activities I just mentioned, a functional prefrontal cortex is very important to daily life.

Unfortunately, this region of the brain—and all its associated tasks—suffers when we are sleep deprived, according to research from the University of California, Berkeley. This can lead to a reduced ability to regulate our emotions. Why? The prefrontal cortex usually helps put emotional experiences into context so that we can respond appropriately, but it becomes dysfunctional with sleep loss, letting the primitive and fearful amygdala (the brain's "fear center") call the shots instead.

Matthew Walker, director of UC Berkeley's Sleep and Neuroimaging Laboratory, said in a release, "It's almost as though, without sleep, the brain had reverted back to more primitive patterns of activity, in that it was unable to put emotional experiences into context and produce controlled, appropriate responses." An amygdala set free of the watchful eye of the prefrontal cortex might be good for an immersive horror movie–watching experience, but it's bad for daily life—especially when it comes to nutrition. Our brains are programmed to seek out sugar, lest they die come winter. With a sleep-deprived prefrontal cortex, say goodbye to your willpower and self-control. If you are prone to overeating or indulging in junk food, just one single night of sleep loss is enough to sidetrack your best efforts at a healthy diet.

The Glymphatic System: Your Brain's Nightly Cleaning Crew

Anatomy textbooks are not updated too often these days. After the advent of the microscope, physiologists were quick to slice, dice, stain, map, and draw every square millimeter of the human body—and within a few short decades, there was seemingly nothing left to explore. So it was a major geek-out moment for biology lovers when Jeffrey Iliff and his team at the University of Rochester discovered what could rightly be called an uncharted organ—the *glymphatic system*. This system forcefully pushes cerebrospinal fluid through the brain while we sleep, providing a free power-wash for our brains every single night.

In the rest of the body, the *lymphatic* system is a physical structure that collects white blood cells and debris and slowly transfers them from the tissues to the bloodstream and lymph nodes (those swollen lumps under your chin when you get a bad cold are activated lymph nodes). But unlike the lymphatic system, the glymphatic system doesn't have a full system of channels and nodes. Because the brain must squeeze into a hard cavity, there isn't room for a large physical network. Instead, the channels of the glymphatic system piggyback on the drainage system of the arteries that supply the brain. In an economical and elegant appropriation of the arterial system for its own uses, the glymphatic system takes over the brain during sleep, causing these ducts to swell up to 60 percent while the neurons themselves shrink to make way for the street-cleaning fluid. And in a final coup, the system commandeers the arteries' pulsations as a way to massage the fluid through the system.

I've already mentioned amyloid, the mischievous protein that clumps together and forms plaques in Alzheimer's disease. We all generate this protein, and the glymphatic system helps to dispose of waste and prevent amyloid accumulation. The system is particularly active during the deep, slow-wave phase of sleep, but unfortunately, today's sleep patterns (and our diets) negatively affect this activity. Thus, poor sleep is associated with greater amounts of amyloid plaque in the brain.[5] We theorize that by optimizing our sleep, we may help these proteins get cleared away before they ever cause problems.

How might we optimize glymphatic clearance? It's still a recently discovered system, and we certainly don't have all the answers yet. However, as discussed in chapter 6, fasting before bed (to reduce circulating insulin) may be one way to encourage the brain and body's custodial duties. Omega-3 fatty acids (abundant in the fat of wild fish and grass-fed beef) were also shown to promote optimal functioning of the glymphatic system.[6] By following the Genius Plan, you will be getting optimal amounts of omega-3 fatty acids. At the end of the day, however, the best way to attain a spotless brain is just to get good sleep, consistently.

There are myriad factors that affect our sleep quality—work stress, family duties, and TV binges that take us into the wee hours. (Who isn't guilty of the occasional Netflix marathon? I certainly am.) But diet may play a significant role here as well: two studies, one published in the *Lancet* and the other in *Nutritional Neuroscience*, both showed that after just two days of a high-carbohydrate, low-fat diet, healthy, normal-weight male subjects spent less time in slow-wave sleep, compared to those on a low-carbohydrate, high-fat diet.[7] Observational research in both men and women has confirmed that higher consumption of sugar and carbohydrates is associated with less time spent in slow-wave sleep. Certain nutrients, on the other hand, may increase sleep quality—higher fiber consumption seems to promote deeper, more cleansing sleep.[8]

If keeping your brain plaque-free wasn't a compelling enough reason for you to reconsider your sleep habits, let me break some more news to you: quality sleep is a precondition for having the willpower to change your other habits, causing hormonal changes that will supercharge your results. Sleep is a keystone for the implementation of all the other changes you're going to make in the Genius Plan.

SLEEP OPTIMIZATION CHEAT SHEET

- **Keep your bedroom cool.** The body likes cooler temperatures to sleep.
- **Take a warm shower or bath before bed.** The drop in body temperature once you step out should signal to your body that it's time to sleep.
- **Use your bed for sleeping (and sex, duh) only.** As soon as you wake up, get out of bed, and don't return to it until you want to go to sleep for the night.
- **Avoid alcohol.** Even though alcohol helps you get to sleep faster, it lessens the amount of time spent in REM sleep, which is your deepest phase of sleep.
- **Avoid nighttime blue light exposure.** Try blue light–blocking glasses (see page 349 for recommendations). Avoid screen expo-

sure and make sure the bulbs in your house are all warm in color temperature.

- **Keep your smartphone far away from your bed.** Anywhere but arm's reach.
- **Keep your room dark.** Even a little bit of light can disrupt sleep. People who slept under a very dim light (10 lux) for just one night had decreased working memory and brain function.[9]
- **Set a caffeine curfew.** Limit caffeine consumption to 4 p.m. at the latest—maybe even earlier if you are a genetically slow metabolizer (a gene testing service like 23andMe can let you know).
- **Eat more fiber and omega-3 fats, and fewer carbs.** Inflammation affects sleep quality, and by-products of fiber consumption (like butyrate) may promote deeper, more rejuvenating sleep.
- **Stop eating at least an hour before bed.** Nocturnal eating can sabotage sleep.[10]
- **Get direct sunlight within twenty minutes of waking, especially during daylight saving time or when traveling.** Bright light helps anchor your body's circadian rhythm, which regulates the natural ebb and flow of your sleep-wake cycles.
- **Use an app as an alarm clock.** Apps like Sleep Cycle wake you only when your sleep has entered one of its lighter phases, preventing that awful feeling of being woken up in the middle of deep REM sleep (your deepest sleep phase).

Hormonal Helpers

Our behaviors are often motivated by our brains, but sometimes they originate in the body. In many ways, willpower is like a marionette puppet, with chemical messengers called *hormones* at the strings. Unlike neurotransmitters, which allow individual neurons to communicate with their next-door neighbors, hormones are long-range messengers, being released in one part of the body and having impact in another. For example, a hormone called *leptin* may come from the fat cells around your belly, directed toward a region in the brain that controls energy expenditure. Or cortisol, secreted by your adrenal glands just above your kidneys, may impact parts of your brain involved in memory.

By understanding the relationship sleep loss and stress have with these master hormone controllers, we may achieve the strongest domain over our willpower—which is to say, we'll rarely have to use it.

Insulin: The Storage Hormone

In chapter 4, I described how excessive insulin might turn our brains into amyloid plaque landfills, but excessive carbohydrate intake is not the only villain in the fight for a plaque-free brain. Sleep is also critical for regulating our hormones, including insulin. Research suggests that even as little as one night of partial sleep deprivation can temporarily increase insulin resistance in an otherwise healthy person.[11]

Short-term sleep restriction has even been shown to increase risk for type 2 diabetes, but there is good news: some of the negative effects of sleep debt appear to be reversed by a weekend of catch-up sleep (around 9.7 hours per night).[12] On the other hand, playing a cat-and-mouse game with sleep is not only a bad habit to get into but a poor strategy for long-term health.

DOCTOR'S NOTE: SLEEP LOSS CAN MAKE YOU—OR KEEP YOU—FAT!

It is impossible to overstate the importance of sleep. In my clinical practice, if a patient comes to me with a goal of weight loss or body recomposition and is sleeping fewer than seven full hours a night, I will say, in not so many words, that she is wasting her money if she's not also committed to improving sleep duration and quality. Recent studies, since replicated, have confirmed that sleep deprivation (which means fewer than six hours of sleep) for a *single night* leads to an unintentional ingestion of an extra 400 to 500 calories the following day, and those additional calories almost always come from carbs. Multiply this by a few nights, and you've got yourself a spare tire in a matter of weeks. Already overweight? Same rules apply: you are seriously hurting your chances of dropping extra weight when sleep deprived.

Ghrelin: The Hunger Hormone

Another hormone affected by sleep is ghrelin. Secreted by the stomach, ghrelin tells your brain when it's time to be hungry. Your ghrelin level increases just before meals or when the stomach is empty and decreases after meals or when the stomach is stretched. This hormone can also impact your behavior: when mice and humans are injected with ghrelin, the number of meals consumed increases.

Ghrelin surges with just a single night of sleep debt.[13] This may be why one night of sleep deprivation will provoke, on average, an excess intake of 400 to 500 calories that day, mostly from carbohydrates, coinciding with increased inflammation, high blood pressure, and cognitive problems.

Aside from getting more sleep, how can we make ghrelin work for us? Eating fewer (but larger) meals throughout the day trains your body to produce less of the hormone. Science has now revealed that the advice to eat small, frequent meals to "stoke the metabolic flame" is bunk: metabolic chamber studies—when volunteers live in a room outfitted with instruments that measure how their bodies use air, food, and water under different conditions—show that whether you eat two meals a day or six your metabolic rate is exactly the same.

This is liberating because the approach of eating fewer and larger meals provides lifestyle flexibility, allows us to feel full, reduces decision fatigue, and helps keep the amount of time insulin is circulating to a minimum. Be aware, though, that as you adjust to fewer and larger meals, it may take at least a few days for your stomach to stop sending out signals saying "It's time to eat!"

Leptin: The Metabolic Throttle Hormone

Sleep can also negatively affect another hormone involved in hunger called *leptin*. Leptin is the "satiety" hormone that helps regulate energy balance by inhibiting hunger, and it plummets with sleep deprivation. Leptin's job is to control energy expenditure through its action on the hypothalamus, the brain's master metabolic regulator. Because leptin is secreted by fat cells, the more fat cells one has, the more circulating leptin. The brain interprets higher levels of leptin as permission to open up the throttle a bit on the rate at which our bodies burn calories—after all, food is seemingly plentiful! But as with insulin, chronically elevated leptin can cause leptin resistance to develop, and the signal of "satiety" and the positive benefits of leptin on metabolism becomes lost.

This is the unfortunate paradox faced by those who lose weight and try to keep it off—they are fighting against the one-two punch of both lower leptin levels and leptin resistance. Low leptin increases hunger while decreasing thyroid activity, sympathetic tone, and energy expenditure in skeletal muscle, adding up to a major metabolic slowdown. Anyone who's undergone massive weight loss understands this system gone awry: a 250-pound person who diets down to 200 pounds usually burns about 300 to 400 fewer calories per day than someone who was 200 pounds to start.

On the other hand, recent studies performed by Harvard obesity researcher David Ludwig suggest that very-low-carb diets can offset some of this metabolic disadvantage to the tune of 100 to 300 calories per day—the equivalent of a daily three-mile run! The great news is that following the protocol outlined in this book will enable you to achieve this metabolic "extra credit."

When we engage in bouts of fasting or very-low-carb diets, our leptin levels are reduced—but this has the benefit of increasing the amount of leptin receptors in the hypothalamus. Thus, by fasting, we can regain leptin sensitivity, and by incorporating periodic high-carb, low-fat "refeeds," we can keep our metabolism revving like a sixties muscle car (for more on this, see the sidebar on hacking leptin).

HACKING LEPTIN TO LOOK BETTER NAKED

Once you are fat adapted, periodic high-carb meals can be a potent way of promoting healthy leptin dynamics. This is because carbohydrate consumption and the insulin secreted as a result are powerful leptin boosters.[14] The corresponding leptin surge messages the hypothalamus to ramp up the body's metabolic engines. This system becomes dysregulated if there is *chronic* high-carbohydrate consumption, which can promote leptin resistance. But when combined with exercise, a weekly higher-carb "refeed" can increase energy expenditure, recalibrate mood, and accelerate fat loss, especially for those who've seen their weight loss stall.

A refeed of 100 to 150 grams of carbs should do the trick. This is still dramatically less than that of a person consuming the Standard American Diet—estimates are that the average Westerner consumes upward of 300 grams of carbohydrates a day. This is also not an excuse to eat junk food. These higher-carb meals should be low in fat, which, if you recall from chapter 2, can potentiate the insulin surge and create a temporary setting of insulin resistance. (Fat can also prevent leptin from crossing the blood-brain barrier.)[15] Great carbs for a refeed include rice (sushi is an excellent refeed food), starchy vegetables like potatoes, or your favorite higher-sugar fruit.

Leptin also plays an important role in cognitive function, which is why keeping it within the normal range (i.e., by not lowering it with prolonged low-calorie dieting or poor sleep, and by topping it off with periodic refeeds) is critical. While leptin's most famous role involves its communication with the hypothalamus, receptors for leptin have also been identified in areas of the brain responsible for emotion, and there is a strong relationship between low levels of leptin and depression and anxiety. From an evolutionary standpoint, this makes quite a bit of sense. Leptin works with insulin to paint a picture for your brain of the state of food availability—and when food is scarce, this likely tells the brain to alter behavior in a way that conserves energy. This can manifest as social withdrawal, an inability to feel pleasure, or a lack of motivation. It should come as no surprise that leptin resistance can contribute to depression. In a recent study, overweight and obese women had significantly increased symptoms of depression and anxiety despite having higher leptin levels than lean controls.[16] For these leptin-resistant women, leptin is present, but the brain cannot sense it.

In terms of overall brain health, leptin is involved in synaptic plasticity in the hippocampus, where it facilitates long-term potentiation—the creation of strong, long-lasting memories. It's been shown to improve memory in rodent models of aging and Alzheimer's disease, and may enhance the clearance of amyloid beta, the protein that accumulates to toxic levels with age. The more you can maintain your sensitivity to leptin, the healthier (and happier) you will be.

Growth Hormone: The Repair and Preserve Hormone

In adults, growth hormone, or GH, is known primarily for its role as a repair hormone. Athletes have been known to use GH for its performance-enhancing qualities, namely its ability to accelerate repair of connective tissue. But GH, which is secreted by the brain's pituitary gland, is also a powerful cognitive modulator, shown to improve many aspects of brain function, including processing speed and mood. In older adults, GH replacement therapy has been shown to boost cognitive function in patients with mild cognitive impairment (pre-dementia that often leads to Alzheimer's disease) and in healthy controls after just five months.[17] But while injecting extra growth hormone is both illegal and potentially dangerous, there are a few hacks at our disposal to naturally boost this powerful chemical.

While growth hormone deficiency manifests as severely stunted growth and stature in children (which led to its initial identification and naming), it has a very different primary role in adults: to preserve lean mass during periods of famine or in a fasted state. One of the best ways to boost growth hormone, therefore, is through intermittent fasting.[18] When one fasts for anywhere from fourteen to sixteen hours onward for a female and sixteen to eighteen hours onward for a male, growth hormone begins to increase. After twenty-four hours of fasting, growth hormone has been reported to shoot up as high as 2,000 percent![19]

Aside from fasting, heat conditioning (sauna use, for example) is a powerful way of boosting growth hormone. In one small but revealing study, two twenty-minute sauna sessions at 80°C (176°F) separated by a thirty-minute cool-off period caused growth hormone levels to double in young male students, while in the same study, two fifteen-minute sessions at 100°C (212°F) resulted in a fivefold increase. Another study that subjected young men to repeated sessions found that two one-hour sessions a day at 80°C (176°F) increased growth hormone by a remarkable sixteenfold, although the increases tapered off after the third day of repeated exposure. As you adapt, spacing out your sessions may be helpful.

As easy as it is to boost growth hormone, it's even easier to deplete it—especially today. Chronic stress is one of the major modern growth hormone fighters, directly at odds with maintaining our precious lean muscle tissue. Carbohydrate consumption immediately turns off growth hormone production, providing an explanation as to why low-calorie diets without carbohydrate restriction can lead to muscle loss concurrent with fat loss.

Finally, getting fewer than seven hours of sleep has been shown to negatively affect growth hormone production. In fact, most of the growth hormone in our bodies is produced during slow-wave sleep, so getting two to three full cycles is critical—shoot for eight hours a night.

Cortisol: The Carpe Diem Hormone

Cortisol, a master circadian regulator, peaks upon waking, creating a temporary catabolic state in the body. Often thought of solely as a stress hormone, cortisol is also instrumental as the "waking" hormone, liberating energy as carbs, fat, and amino acids for use in the early daytime hours. When insulin and cortisol are both present at the same time (i.e., after a carbohydrate-rich breakfast), cortisol's fat-burning effect will be shut down, and it will only exert its catabolic effect on your muscles—clearly, not a desirable scenario.

While skipping an early breakfast may help cortisol fulfill its job, should you choose to eat a morning meal it should consist solely of fat and protein and fibrous veggies—*not* carbohydrates. This is contrary to the popular dogma of starting the day with a hearty bowl of oatmeal or cereal—to say nothing of the bagels, muffins, pancakes, pastries, and more that are so commonly consumed in the morning.

CORTISOL'S DARK SIDE

National Geographic journalist Dan Buettner has discovered and studied the five places in the world—dubbed Blue Zones—where people live the longest. The lifestyles of people in these zones provide examples that we can use to form hypotheses about what promotes healthy aging. For example, many of these communities build nonnegotiable intermissions from work into their days—and I'm not just talking about a lunch break. "The world's longest-lived people have routines to shed that stress," Buettner writes in *The Blue Zones*. He continues:

> The Okinawans call it *ikigai* and the Nicoyans [of Costa Rica] call it *plan de vida*; for both it translates to "why I wake up in the morning." In all Blue Zones people had something to live for beyond just work. Research has shown that knowing your sense of purpose is worth up to seven years of extra life expectancy.

Unless we develop effective ways of defusing stress (which, let's face it, is an unavoidable aspect of twenty-first-century living), cortisol can become elevated for long periods, resulting in some serious physiological consequences.

But before we get to that, let's define what chronic stress is and isn't. Chronic stress is *not* what you feel when giving an occasional presentation, suffering through a move, or the chance traffic jam when you're already late. The forms that chronic stress usually take are the following (and make a mental note if any of these seem familiar):

- Showing up every day to a job that you hate
- Prolonged financial hardship
- Having to work under a boss you dislike
- Being stuck in a long-term relationship gone sour
- Having a bully at school
- Military duty
- Chronic noise exposure

- A stressful daily commute to and from work
- Medical school (—Dr. Paul)

Unpleasant, protracted, and a recent invention in evolutionary terms, this type of chronic stress activates the amygdala, that primitive survival region associated with fear. Its job is to kick off a cascade of biochemical processes that were initially meant to help scuttle us from harm's way when we were confronted by a physical threat—say, a lion charging toward us on the savanna. Imagine this scenario: you are a hunter-gatherer going about your day, peacefully foraging for berries under the hot East African sun. Suddenly, a lion appears in your periphery—let's pretend for the sake of the story this lion's name is Mufasa. Mufasa hasn't eaten in days, nor has his hungry cub back at the pride (let's call him Simba). Mufasa sees you as the perfect meal to break his fast and feed his cub—rich in protein, calories, and omega-3s—and begins hurtling toward you at full speed.

At this moment, your amygdala, which is essentially your brain's lookout deck, kicks your sympathetic nervous response into motion, priming your body for action. The amygdala activates something called the *hypothalamic-pituitary-adrenal (HPA) axis*, causing the adrenal glands to secrete cortisol and epinephrine (also known as adrenaline), and suddenly, your mellow day of flower picking becomes an all-out sprint for your life.

THE HPA AXIS—THE STRESS RESPONSE SWITCHBOARD

Once initiated, the HPA axis begins in the brain structure known as the hypothalamus—the *H* in *HPA*. One of the most important functions of the hypothalamus (aside from its role as a metabolic master controller) is to link the brain to the body's hormone system via the pituitary gland. The hypothalamus sends out some corticotropin-releasing hormone (CRH) to the pituitary gland—the meat in the HPA sandwich. After receiving word of turmoil from the hypothalamus, the pituitary gland then secretes something called *adrenocorticotropic hormone* (ACTH) into circulation. (Recall that hormones are long-range messengers, which differ from neurotransmitters, which act neuron-to-neuron.) Now in circulation, ACTH acts on the adrenal glands, which are perched above the kidneys. This causes a surge in cortisol and epinephrine.

HPA Axis: **H**ypothalamus ➤ **P**ituitary gland ➤ **A**drenal glands

Amygdala ➤ Hypothalamus (corticotropin-releasing hormone [CRH]) ➤ Pituitary gland (adrenocorticotropic hormone [ACTH]) ➤ Adrenal glands (Cortisol) ➤ Circulation

---

The cortisol and adrenaline that are now coursing through your body have several effects on your physiology. For one, heart rate and blood pressure go way up. Pupils dilate. Salivary secretion halts and digestion slows down (digestion is a relatively labor-intensive process, and running from Mufasa is no time to be using precious resources on nutrient absorption). In fact, blood leaves the digestive area, rerouting to more important locations, like your muscles. Blood sugar is released from the liver, and the parts of the body that are inessential to getting you out of harm's way become resistant to insulin, making sure that your muscles can get all the glucose that they need. The immune system becomes suppressed, and blood itself becomes more viscous as platelets (the type of blood cells involved in clotting) begin to aggregate as a cautionary measure in case of blood loss.

The odds of being chased by a lion are pretty slim these days. If you're lucky, true threats of a physical nature aren't a common occurrence. But while our sources of stress have evolved, our response to them has not. So when you have an argument with a coworker, run to catch your subway only to be left on the platform as it screeches away, or get frightened by the air horn of an eighteen-wheeler as it lurches alongside you in traffic, the same domino effect begins in your body. When you have multiple stressful stimuli back to back, your body's response can create serious problems. This is why stress is such a vicious, indiscriminate killer—the chronic activation of this antiquated system, once lifesaving, now promotes inflammation, elevated blood sugar, insulin resistance, nutrient deficiencies, increased gut permeability, and more. But chronic stress *plus* carbs? That's a recipe for disaster.

At this point, it may not surprise you to know that as our waists grow, our brains shrink.[21] We've already covered many factors that might explain this startling observation, except one: chronically elevated cortisol due to stress.

Ever see a person with a bulging midsection but surprisingly skinny arms and legs? This is the picture of chronic stress. It's completely different from your run-of-the-mill obesity, where everything—legs, arms, butt—is blown up to comparable proportions. This is because deep abdominal fat, the kind that wraps itself around your heart, liver, and other major organs, not only receives more blood but has four times more cortisol receptors than subcutaneous fat (the fat that you can "pinch" below the skin).[22] When cortisol is elevated, any carbohydrate intake will immediately promote fat storage, and most likely as the deep abdominal fat called *visceral fat*, which is the most dangerous and inflammatory kind of fat. This makes concentrated carbohydrate consumption uniquely damaging to a stressed-out person. (This is another reason why eating carbs first thing in the morning, when cortisol is naturally at its peak, is a bad idea.)

If you are experiencing a bout of stress, the reaction should be twofold: first, deal with that stress, and second, keep concentrated sources of glucose and fructose especially low. Here are some other important stress-abating tips:

- **Meditate, don't medicate.** Meditation can be intimidating for first-timers, but it's worth getting comfortable with it. One small Thai study of stressed-out medical students found that four days of meditation reduced cortisol by 20 percent.[23]
- **Spend more time outdoors.** We've lost touch with nature, but merely seeing greenery mitigates the physiological response to stress and improves cognitive function.[24] Being in nature can also help reduce depressive thoughts and even boost BDNF.[25]
- **Exercise smarter.** Alternate between "low and slow" aerobic sessions (a bike ride or a hike in nature) and more intense bursts. Chronic medium-intensity cardio sessions (running on a treadmill for forty-five minutes, for example) can actually *increase* cortisol. We'll see more on this in chapter 10.
- **Have somebody give you a massage (or pay for one—never a bad investment!).** A 2010 study out of Cedars-Sinai Medical Center in Los Angeles found that five weeks of Swedish massage significantly reduced serum cortisol compared to controls who underwent only "light touching."
- **Practice deep breathing.** Simple yet effective. Exhaling activates the parasympathetic nervous system, responsible for the body's "rest-and-digest" processes.

It's been known for some time that a chronic elevation of cortisol compromises the brain's supply of BDNF and can atrophy vulnerable structures like the hippocampus, even causing dendrites (the physical correlates of memories) to recede.[26] This reinforces the negative aspects of stress, since the hippocampus normally "vetoes" inappropriate stress responses. Repeated stress therefore hurts your ability to control stress, and this has been borne out in the research. In mice that were subjected to chronic "social defeat"—the equivalent of having a bully in the cage with them—their memories suffered significantly. If the neural pathways created by learning are akin to an ever-expanding set of train tracks, it seemed as if those mice under duress were having trouble laying down new tracks.

Recent research has also highlighted new mechanisms by which chronic stress can impair your long-term brain health. Chronic stress was shown to actually activate the immune system of the brain, producing inflammation almost as if the brain were responding to stress as an infection. Inflammation is the cornerstone of many neurodegenerative diseases, as I've mentioned throughout this book. But recently, chronic exposure to stress hormones was connected to the hallmark plaque that characterizes Alzheimer's disease. Long-term administration of cortisol was shown to reduce levels of insulin-degrading enzyme (IDE) in the brains of monkeys.[27] IDE is responsible for breaking down insulin in the brain as well as amyloid beta—the protein that clumps together to form the plaques characteristic of Alzheimer's disease. (Revisit page 100 for a primer on IDE.)

As you can see, chronic stress is a major threat to our cognitive health. But not all stress is created equal! In the following chapter, we will see how one particular kind of stress may be your brain's best friend.

FIELD NOTES

- Sleep is sacred—it keeps your hormones healthy, helps your brain better regulate your emotions, and can even help you lose weight.
- Sleep is also when your brain cleans itself, offering a free power-wash every night thanks to the newly discovered glymphatic system.
- We can optimize sleep and the glymphatic system with a high-fiber, low-carbohydrate diet.

- Fasting can dramatically increase growth hormone, which protects lean mass.
- For a fat-adapted low-carb dieter, the occasional high-carb, low-fat "cheat day" can help to increase leptin levels, which will encourage fat burning and enhance mood.
- Stress management is paramount to health—chronic stress grows our waists, shrinks our brains, and creates performance-zapping inflammation.

GENIUS FOOD #9

WILD SALMON

The consumption of wild fish has been long associated with reduced risk for cardiovascular disease, cancer, and even all-cause mortality, but what about its impact on the brain? I'm glad you asked, because consumers of wild fish exhibit superior cognitive aging and better memory function, and even possess bigger brains![1] In a recent study, cognitively normal older people who ate seafood (including fish, shrimp, crab, or lobster) more than once per week had reduced decline of verbal memory and slower rates of decline in a test of perceptual speed over five years compared to people who ate less than one serving per week. The protective association of seafood was even stronger among individuals with the common Alzheimer's risk gene, *ApoE4*.

The king of these fishes is wild salmon, which is low in mercury and a rich source of both EPA and DHA omega-3 fats and a powerful carotenoid called *astaxanthin*. Derived from krill (the main food supply of wild salmon), this carotenoid is added into the diets of farm-raised salmon to give the characteristic "pink" to their color, but it is far more abundant in salmon that is wild (hence their richer color). Astaxanthin is beneficial to your entire body and can help do the following:

- Boost cognitive function and promote neurogenesis
- Protect the skin from sun damage and enhance skin appearance
- Protect the eyes, reducing inflammation

- Convert blood lipids to a more cardioprotective profile
- Provide potent antioxidant effects and free radical scavenging

Some of these benefits appear to be facilitated by astaxanthin's unique molecular structure, which allows it to shield cell membranes from oxidative stress. On top of that, it has also been shown to "switch on" genes that protect against DNA damage and the stresses of aging, such as the *FOX03* longevity pathway, described on page 98. Shrimp, crab, and lobster are also high in astaxanthin and are good options when you're looking for a little variety in your wild fish consumption.

How to use: Broil, pan-sear, poach, or eat raw (if sashimi-grade).

Pro tip: All types of fatty fish, including sardines, herring, mackerel, and anchovies, are good alternatives. I will often travel with canned sardines to serve as a quick snack or to add to a meal, and I have included them in my Better Brain Bowl (page 333). Just make sure any canned fish is packed in olive oil (ideally extra-virgin) or water only.

CHAPTER 10

THE VIRTUES OF STRESS (OR, HOW TO BECOME A MORE ROBUST ORGANISM)

Consider that Mother Nature is not just "safe." It is aggressive in destroying and replacing, in selecting and reshuffling. When it comes to random events, "robust" is certainly not good enough. In the long run everything with the most minute vulnerability breaks, given the ruthlessness of time—yet our planet has been around for perhaps four billion years and, convincingly, robustness can't just be it: you need perfect robustness for a crack not to end up crashing the system. Given the unattainability of perfect robustness, we need a mechanism by which the system regenerates itself continuously by using, rather than suffering from, random events, unpredictable shocks, stressors, and volatility.

—NASSIM NICHOLAS TALEB,
ANTIFRAGILE: THINGS THAT GAIN FROM DISORDER

In fewer words:

That which does not kill us makes us stronger.

—FRIEDRICH NIETZSCHE

Finding stagnation in the universe is a difficult task. It simply doesn't exist. Celestial bodies are being either slowly created or slowly destroyed. Here on Earth, stagnation is associated with rot and decay, like a pond that has lost its inflow. For our brains, it is a death sentence.

Like all matter in the universe, we are subject to the second law of thermodynamics: entropy. This fundamental law of physics states that all systems, over time, decline from states of higher complexity to lower complexity. This slow transition from order to disorder is what occurs to stars, planets, and entire galaxies, and is also what happens to us during the aging process.

At first, however, human life seems to defy this law, in the profound regenerative abilities that a child exhibits. Children don't often develop cardiovascular disease (stop the presses: signs of it are creeping up in children as young as eight due to the ravages of the Standard American Diet). They don't get dementia, and almost 90 percent of pediatric cancer cases are curable. These "superhuman" abilities are seemingly lost during our elder years.

What if we could turn back time and regain the level of resilience we've all exhibited in youth? To "rage against the dying of the light," so to speak. I'm here to tell you that it may be possible, in a form that has for a long time been demonized in mainstream and medical literature alike: stress, the antidote to stagnation.

Now before you throw your hands up in confusion, let me clarify: there are two kinds of stress. There is chronic stress—the kind that comes from a bad job, sour relationship, prolonged financial hardship, or even what my friend fitness author and mega-athlete Mark Sisson calls "chronic cardio" (discussed momentarily). This kind of stress accelerates entropy and decay. It leads to prolonged elevation of the hormone cortisol, which can rob our muscles of strength and redistribute our body fat to our bellies, cause important parts of the brain to atrophy, and even accelerate the aging process.

Acute (or temporary) stress is a different beast entirely and may be one of our most powerful weapons in the fight against entropy. This form of stress may take many shapes. It may be the mental stress that one endures when learning to play an instrument, engaging in a particularly challenging and lifelike video game, or sitting through a tough lecture. It can also be physical stress, in the form of exercise, brief bouts of fasting, extreme temperatures, or even certain types of "stressful" foods.

Hormesis, one of my favorite biological principles, is the mechanism by which small doses of stress from, say, a tough workout, a good sweat in the sauna, or even temporary calorie restriction (which we call intermittent fasting) can promote more efficient cells and greater long-term health. While large doses of a particular stressor might harm you, small doses actually cause your cells to adapt and grow stronger. The following pages will explore how you can leverage the power of hormesis to supercharge your cognition and help you live stronger for longer.

Movement

Now, here, you see, it takes all the running you can do, to keep in the same place. If you want to get somewhere else, you must run at least twice as fast as that!

—**THE RED QUEEN, LEWIS CARROLL'S *THROUGH THE LOOKING-GLASS***

I've always been terrible at sports. The few summers my parents were brave enough to send me off to camp, I abstained from playing games like football, soccer, and dodgeball and gravitated to archery, rocketry, and ceramics instead. (When it came time for swimming, I was always too shy to take off my shirt, an insecurity that I've thankfully outgrown.) In high school, rather than join the basketball team like many of my peers, I was drawn to computer programming.

I became interested in the gym only when I learned exercise could manifest as a stronger or leaner body. I began to see both food and exercise as "code" to speak to my biological programming.

In retrospect, I realize that many of the same feedback loops that drew me to programming were also present in fitness, including the ability to simplify my routine and debug problems. These feedback loops provided enough dopamine hits to hook a shy and introverted sixteen-year-old computer programmer (and the increased attention from my female classmates wasn't bad either).

That exercise is one of the greatest known means of enhancing cognitive function, mood, and neuroplasticity really shouldn't come as a surprise. When it all boils down, we're a species made to move. And yet, along with our diets, our lifestyles have undergone a dramatic shift. We used to roam thousands of miles on foot as hunter-gatherers, and when we weren't walking or hiking, we were running—not sitting at desks, on trains, or in cars stuck in traffic.

Just how adapted to movement are we? Recently analyzed fossils of aboriginal footprints show a stride indicating that our ancestors were, on *average*, at least as fast as Usain Bolt, the Olympic sprinting champion. Other signs are evident in our bodies: we're excellent at dissipating heat through sweat. We have long legs, large knees, and the springlike Achilles tendon, which despite its name is one of the strongest soft structures in the animal kingdom. And with relatively voluminous backsides and a large percentage of fatigue-resistant, slow-twitch muscle fibers, we may also be the endurance athletes of said kingdom.

Today, however, we grab our lunches to go so that we can sit and eat in isolation at our desks. We're mostly stationary throughout our workday and during our commute. And then we sit on the couch and binge-watch hours of television. Research over the past few years has validated the notion that chronic sitting is bad for us. So bad, in fact, that some experts have even gone so far as to call sitting the new smoking. While this may sound hyperbolic, excessive sitting *has* been linked to early mortality, representing nearly 4 percent of annual deaths worldwide.[1] This association shrinks dramatically with the addition of even just a little bit of extra movement in the day:[2] one study from the University of Utah found that just two minutes of walking for every hour spent sitting dramatically reduced (by 33 percent) risk of early death, while a study out of the University of Cambridge found that an hour of moderate-intensity exercise a day eliminated it completely.[3]

For the brain, exercise might as well be considered a panacea, continually validated by research trials in both the cognitively healthy and impaired. It acts as both medicine and a tonic, coating our vulnerable organ in a chemical cocktail of "smart" molecules ranging from powerful antioxidants to nerve growth factors. And after reading this section, you'll know exactly how to implement exercise for maximum cognitive gain.

The Brain Grower

Okay, so you're sold on exercise. Where do we start?

There are two main energy systems that you can train—aerobic and anaerobic. For simplicity's sake, aerobic exercise is akin to a long bike ride or hike, while anaerobic exercise tends to include weight lifting and sprinting. Think of the former as fat and oxygen burning and the latter as sugar burning.

Aerobic training gets your heart rate up and can be sustained for a long period of time. The vast majority of the day you are functioning in a state of aerobic respiration. Aerobic *exercise* simply increases the intensity and demand on your metabolism, but under similar metabolic conditions.

AEROBIC EXERCISE

Low and Slow!

- Hiking
- Bike ride

- Long brisk walk
- Light yoga

All forms of exercise help to increase blood flow to the brain, pushing desperately needed oxygen and nutrients to our biological control centers, but aerobic exercise in particular has been found to be one of the best known means of boosting brain-derived neurotrophic factor, or BDNF. I've used phrases like "Miracle-Gro for the brain" and "the brain's ultimate fertilizer" throughout this book to convey the powerful effect BDNF has in terms of promoting neuroplasticity and protecting your brain cells, but I admit these may still seem like abstract concepts. (Sadly, we can't flex our hippocampi in the mirror.) If you had access to an MRI machine, however, you would be able to see the profound growth that BDNF promotes.

A seminal study published in 2011 gave scientists the opportunity to do just this.[4] It involved 120 cognitively healthy adult subjects, half of whom regularly performed an aerobic exercise routine three times per week over the course of one year. Using an MRI, scientists saw that aerobic exercise increased the size of participants' hippocampi by 2 percent over what it was at the beginning of the study. Now before you scoff at what may seem like a very modest increase, you should know that the hippocampus typically *loses* volume at a rate of about 1 to 2 percent each year after the fifth decade of life. And in fact, this happened in the control group—their scans showed that the same degree of brain volume was actually lost. As noted by the researchers, aerobic exercise essentially turned back the clock on the hippocampus, the brain's memory-formation center, by one to two years. As I write this, there's no drug in the known universe that wields this degree of power. If that weren't exciting enough, the leap in size seen in the brains of the exercising group coincided with a performance boost in the type of memory used to navigate familiar places.

CLOTHE YOUR BRAIN IN *KLOTHO*

Klotho is a longevity protein named after Clotho, the Fate from Greek mythology who is known for spinning the thread of life. Were Clotho real, she might be pleased to know of her affiliation with this "aging suppressor" tasked with, among other things, making tighter, better connections at the synapse, the microscopic junctions where all neural processes occur.

Independent of its effect on healthy brain aging, Klotho has a marked effect on cognition.[5] About one in five lucky people have gene variants that cause them to create more of this protein, and a recent study found that those who had this gene tested on average six points higher on a test of broad cognitive domains, including language, executive function, visual and spatial intelligence, and learning and memory. While you may be tempted to exclaim, "See, it's all in your genes!" the good news is that Klotho can be boosted with aerobic exercise. What's more, Klotho expression, like BDNF, is thought to be dependent on fitness level, so the more exercise you perform (and the fitter you become), the more just a single bout of exercise will raise levels of Klotho.[6]

But beefing up your hippocampus doesn't just protect you from aging. While the hippocampus is one of the first parts of the brain to be attacked in Alzheimer's disease, it is also highly vulnerable to the insults of chronic stress. Chronically elevated cortisol, a consequence of the body's "fight-or-flight" system being overstimulated, can damage the hippocampus. This creates a negative feedback loop, as it is the hippocampus that largely dictates how calmly (or frantically) the brain will respond to a given event. This is because regions of the brain involved in fear and emotion "consult" with the hippocampus to determine how best to respond. As research has shown, by reinforcing this memory structure with exercise, we powerfully augment our brains to become more resilient to psychological stress.

EXERCISE: THE DEMENTIA SLAYER?

One gene in particular, the *ApoE4* allele, continues to come up throughout this book. While far from a sentence to develop dementia, it is the only well-defined Alzheimer's risk gene, and having one or two copies does increase the likelihood that a person will develop cognitive decline. Research suggests that exercise can negate some of the gene's observed influences on the brain. It does so in part by "normalizing" brain glucose metabolism, which is reduced in *ApoE4* carriers (discussed in chapter 6), and by reducing plaque buildup, which seems to be accelerated in carriers. Interestingly, the *ApoE4* allele is considered the "ancestral" (i.e., oldest) variant of the *ApoE* gene, having emerged at a time when we had to chase our food. Its negative association with modern disease may be merely a consequence of our recent transition to relative inactivity, amplified by our industrially mangled diets.

If neurological decline is one consequence of inactivity, might becoming more active actually *reverse* cognitive impairment? Researchers sought to answer this question in a 2013 study, ultimately finding that sedentary people with mild cognitive impairment (MCI) improved their memory and the efficiency of their brain cells after just three months of regular exercise.[7] The study also included a group of cognitively normal people who saw similar benefits. What's more, the subjects improved their cardiorespiratory fitness by just 10 percent, suggesting big cognitive gains for a relatively small boost in fitness.

A follow-up study published in 2015 found that for both healthy older people and patients with MCI, exercise increased the size of the *cortex*, the outer layer of the brain that shrinks dramatically in late-stage Alzheimer's disease. In a very simplistic metaphor, the cortex can be considered the brain's hard drive, where memories are stored after being input by the hippocampus, the brain's keyboard. Participants who showed the greatest improvements in their fitness had the most growth in the cortical layer. Studies like this are key because MCI is considered a critical stage of cognitive decline that can lead to Alzheimer's disease or other forms of dementia.

The Metabolic Enhancer

No man has the right to be an amateur in the matter of physical training. It is a shame for a man to grow old without seeing the beauty and strength of which his body is capable.

—**SOCRATES, CIRCA 400 BC**

While aerobic activity is the chief way to fortify the brain with new brain cells, anaerobic exercise is the best way to keep those cells healthy and metabolically efficient.

Unlike aerobic exercise, which can potentially be sustained for hours (particularly low- to moderate-intensity varieties), anaerobic modes of metabolism are experienced in bursts, achieved through physical activity that is performed at a much higher intensity (and therefore impossible to sustain). This may include sprinting at near-maximal effort for ten to twenty (or even thirty) seconds, taking a break, and repeating the process. Resistance training—weight lifting, for example—is also anaerobic. While everybody's anaerobic threshold will be different, the principle is the same:

by momentarily overloading your body, you provide a powerful stimulus for your cells to adapt, grow stronger, and become more efficient.

ANAEROBIC EXERCISE

Hard and Fast!

- All types of "burst" exercise (such as sprints, vigorous biking, rowing, battle ropes)
- Weight lifting
- Steep hill climbing
- Interval training
- Isometrics
- Intense yoga

One visible benefit of anaerobic exercise is that over time, your muscles grow. This is particularly beneficial for weight maintenance. While anaerobic exercise itself burns far fewer calories than aerobic exercise (the long run on a treadmill, for example), creating even just a little more muscle is beneficial for long-term weight loss. This is because the more muscle you have on your body, the higher work capacity you will have, the more high-intensity activity you can sustain, and the more calories you can soak up without storing them as body fat. Every time you hit the *lactate threshold* in your workouts, which is when your muscles start burning and trembling as you approach failure, you are emptying the stored carbs in the muscle and turning your body into an energy sponge. This means that when you consume a starch like rice or a sweet potato, the carbs become more likely to be shuttled into your muscle cells where they stay, waiting to power your next workout. And increasing muscle mass means more calories burned to fuel those muscles, even when you're just waiting in line to check out at the supermarket.

Pushing yourself to your physiological limits, however, confers benefits that extend far beyond bathing suit season. At the microscopic level, your mitochondria, the organelles that create cellular energy, feel the burden of increased demand. This is in part due to the surge in production of reactive oxygen species (ROS), a normal by-product of metabolism. You may also know of ROS by another name—free radicals. Under normal circumstances, we want to keep these free radicals to a minimum, but in the case of exercise, their increase acts as a powerful signaling mechanism, setting off a cascade of events at the genetic and cellular levels meant to protect us—and enhance our resilience to future stress.

DOCTOR'S NOTE: OBEY YOUR BIOLOGICAL IMPERATIVES

Feeling the occasional bout of melancholy is a perfectly normal, and probably even healthy, aspect of the human condition. But if melancholia turns to negative self-talk, remember: don't judge your thought content or your mood unless you've been working out regularly. If you didn't walk your dog or let him out to play or run around every day, it would be considered animal abuse, and yet we seem to think it's okay to not move ourselves. Exercise should be the last thing to be abandoned when you're feeling busy or overwhelmed, not the first. When tested head-to-head against multiple antidepressants, three days a week of moderate exercise was determined to be *equally effective* as the pharmaceuticals, with the pleasant side effect of having zero side effects! Treat yourself at least as well as you do your puppy—you deserve it.

One enzyme that becomes activated during anaerobic activity is called *adenosine monophosphate-activated protein kinase*, or AMPK. Known as the metabolic "master switch," AMPK acts like a tuning fork to your mitochondria, increasing fat burning and glucose uptake, and turning on waste disposal machinery to clear up cellular junk (this includes recycling old and damaged mitochondria). Activating AMPK is such a powerful means of enhancing cellular vigor that the diabetes drug metformin, which stimulates AMPK, is now being studied for its potential as a geroprotective, or anti-aging, agent. (Preliminary research suggests it may improve symptoms in early Alzheimer's disease and help reduce one's risk for developing it.) But you don't need drugs and their potential side effects to activate AMPK, because short bursts of intense exercise do it as effectively.

One of the most important ways that AMPK enhances your metabolism is by stimulating the creation of more mitochondria, a process called *mitochondrial biogenesis*. Having more mitochondria is generally considered a good thing, and we know this because chronic disuse of muscle, sedentary behavior, and aging each independently result in a decline in mitochondrial content and function.

By creating new mitochondria in our muscle, we improve fitness and metabolic health—including sensitivity to insulin. This is why stimulating AMPK with anaerobic exercise (like weight training and sprinting) is one of the best known means of reversing insulin resistance, along with dietary change.* But AMPK doesn't just spur this dramatic increase in mitochondria in our muscle tissue. It also does this in our fat cells, a process called *browning*. Once believed to be present only in newborns, brown fat is mitochondria-rich fat tissue whose main purpose is to burn calories to warm us up when we get a little chilly (a process called *thermogenesis*).

* Many obese and insulin-resistant patients are told to focus their energy on "doing more cardio" to lose weight, which ignores what would be the more appropriate aim of gaining more muscle to reestablish insulin sensitivity.

Mitochondrial biogenesis, spurred by exercise, was also shown to occur in brain cells in animal research.[8] This has obvious implications, not only with regard to fighting mental fatigue and cognitive aging, but also with respect to neurodegenerative diseases that involve mitochondrial dysfunction, including Alzheimer's disease, Parkinson's disease, and ALS. This may be why a large King's College London twin study showed a strong link between leg strength (involving the largest muscles in your body) and brain volume, with decreased cognitive aging across ten years.[9]

All this taken together is why anaerobic exercise is a vital part of the brain health and cognitive optimization equation. Arthur Weltman, who heads the exercise physiology laboratory at the University of Virginia in Charlottesville, perhaps said it best in an interview with the website ScienceNews: "In order for physiological systems to adapt, they need to be overloaded." Whether that means hitting the weight room and "lifting heavy things," pushing yourself to your limits on a stationary bike for a few moments (rinse and repeat), or adding some all-out sprinting to your cardio routine, including anaerobic exercise in your routine is a major opportunity to optimize your cognitive function.

HIGH-DOSE ANTIOXIDANTS—A CELLULAR CRUTCH?

The call to your cellular machinery to grow stronger is signaled by a temporary increase in free radical–mediated stress, brought on by exercise. Take this pressure away and exercise becomes less effective. This was made evident in a University of Valencia trial in which high doses of an antioxidant, vitamin C, were given to athletes just prior to training. Not only was their performance negatively affected as a result, but many of the aforementioned benefits of exercise—increasing antioxidant coverage and mitochondrial biogenesis—were blocked.[10]

Studies like this highlight a potential negative effect of high-dose vitamin supplementation, which can indiscriminately block the stimulus our bodies need to grow stronger. For this reason, I do not recommend excessive vitamin supplements—instead, a wiser approach is to naturally stimulate the body's own, far more potent antioxidant compounds with exercise and foods such as avocados, berries, kale, broccoli, and dark chocolate (conveniently, all qualifying Genius Foods).

How to Get the Most out of Exercise

As you can see, both aerobic and anaerobic exercise provide unique benefits to the brain and body that extend well beyond calories burned. But how much effort do you need to put in for maximum benefit? Surprisingly, a lot less than you would expect. The latest research suggests our aerobic workouts should be longer and slower, while our anaerobic workouts should be shorter and more intense. What we want to definitely avoid is "chronic cardio," or sustained high-output training, such as a hard forty-five-minute run multiple times per week. There is a peak point where we stress the body enough to stimulate adaptation, but more is not necessarily better. For example, longtime marathoners lose lean mass, drop their testosterone levels, develop increased intestinal permeability, and can even develop scarring of the cardiac muscle and electrical conduction system, leading to dangerous and life-threatening arrhythmias, not to mention the wear and tear on their joints from thousands of steps.

So what's the sweet spot? Essentially, instead of a grinding 45-minute run with a grimace on your face the entire way, you are better off doing a 90- to 120-minute hike, smiling and conducting a conversation the whole way. The low, slow movement, like hiking, will also help move lymphatic fluid around your body, develop your capillary beds, and keep your joints healthy. Otherwise, short sprints at 90 to 95 percent of maximum effort have been shown to have the same improvements in cardiorespiratory fitness and endurance in 20 percent of the time compared to steady-state cardio!

A proper exercise routine should bleed into one's overall lifestyle, including aerobic work (such as long walks, hikes, and biking to and from work) and concentrated anaerobic exertion. This way you maximize neuroplasticity through BDNF with aerobic exercise while achieving the metabolism-fortifying effects of anaerobic exercise.

FAQ: I'm a woman. Won't weight lifting make me bulky? I gain muscle really easily.

A: Muscle is hard earned and gains come over a period of years, not weeks. And if you're afraid of looking "jacked"—trust us, you don't get ripped by accident. Dr. Paul and I have both been trying to get jacked **for the last twenty years**. On top of that, most females simply don't have the hormonal profile to "get huge" without illicit substances.

As long as you're gaining muscle and not overeating or gaining significant weight, your body composition will improve. Your waistline will shrink, and your arms will, yes, shrink even though they are stronger. Regular weight lifting also gives you a buffer for any carbs that sneak in here and there. Just remember: strong is the new skinny.

It's also possible to incorporate aspects of both forms of exercise in the same workout. For example, if you enjoy weight training but dislike running, you can add an aerobic aspect to your workout simply by shortening the amount of rest in between sets. Alternatively, you may choose to do anaerobic workouts on some days of the week and aerobic on others. Whatever you choose to do, stick with what you enjoy and make sure to vary levels of intensity. Also, taking one to two days off per week for rest ensures that you do not overtrain, which can have deleterious effects.

A typical week might look something like this:

Monday	Thursday
Resistance training	Resistance training
Option A: Squats, deadlifts, kettlebell swings, bench press, push-ups, dips, pull-ups, rows, lunges	Option A: Squats, deadlifts, kettlebell swings, bench press, push-ups, dips, pull-ups, rows, lunges
Option B: "Push" exercises: bench press, incline press, overhead press, battle ropes	Option B: "Pull" exercises: Pull-ups, rows, curls, stiff-legged deadlifts, battle ropes
Tuesday	**Friday**
Yoga	Yoga
Wednesday	**Saturday**
A bicycle ride to and from work	Sprints in the park
	Sunday
	A long hike or walk

Heat Conditioning

If one culture can be credited for singlehandedly bringing hyperthermic (i.e., heat) conditioning into the mainstream, it may very well be the Finnish. Sauna use is an integral part of daily life in Finland, where there is on average one sauna for every household![11] Some of these saunas are built in the unlikeliest of places—in an abandoned phone booth, on a boat, or in an immobilized Airstream. Many of these are caught on film in the quirky documentary *Steam of Life*, chronicling this musty national pastime. Elsewhere in the world, though, saunas tend to be found in spas and higher-quality gyms.

While you may be inclined to write saunas off as mere recreation with a *schvitz*, science is beginning to validate their use as a powerful health-modulating activity. Recent research has shown both mechanistically and observationally that hyperthermic therapy gives your brain a potent workout and might play a powerful role in protecting it from aging.

Heat Shock Proteins: A Protein Bodyguard

Sitting in a hot sauna imposes a certain kind of stress on your body called—wait for it—*heat stress*. The highly adaptable human body, forged in the climate of East Africa, knows that heat can kill you and, as a result, takes precautionary measures to protect itself. One such protective measure includes activating *heat shock proteins*, or HSPs. As their name implies, heat is the primary variable to get HSPs going, although these proteins are also activated by exercise and cold temperatures.

HSPs act to guard other proteins by protecting them from "misfolding," as the consequences of protein misfolding are widespread. The unique three-dimensional configurations of proteins help them to be recognized by various receptors, providing "lock-and-key" functionality that allows them to perform many important jobs around the body. Proteins that become disfigured via misfolding are not only less effective; they become strangers to the immune system, which could evoke an autoimmune response.

The misfolding of proteins is also directly implicated in a few diseases that you may be familiar with: Alzheimer's, Parkinson's, and Lewy body dementia. All of these are classified as "proteopathic" diseases, meaning that proteins become misfolded and clump together into plaques; in Alzheimer's it's the amyloid beta protein, whereas the alpha-synuclein protein is involved in Parkinson's and Lewy body dementia. But these plaques form in everyone, not just patients diagnosed with dementia, and doing what we can to prevent the formation of these plaques is worthwhile—especially if it's as simple as sitting in a sauna.

One study, published in 2016 in the journal *Age and Ageing*, provided first-of-its-kind population-level evidence that regular sauna use may indeed help save our brains from decline. Involving more than two thousand people who were followed over twenty years, the study showed that sauna use four to seven times per week led to a 65 percent reduced risk of developing Alzheimer's disease or other dementia, even after controlling for other variables like type 2 diabetes, socioeconomic status, and cardiovascular risk factors.

Brain-Boostin' BDNF

Who doesn't appreciate a free lunch now and then? While exercise is an incredible way to nourish your brain with BDNF, heat stress (from post-workout sauna use, for example) may push BDNF beyond what is achieved by exercise alone.[12]

To explore the synergy between exercise and ambient temperature, University of Houston scientists studied the neural effect that occurred when mice ran in cold or hot temperatures (either 40°F or 99.5°F).[13] In both settings, mice generated a greater number of neurons in the hippocampus, despite running much shorter distances than those in the control group that ran at room temperature. What this suggests is that performing brief periods of exercise in either cold or hot ambient temperature may accelerate the brain benefits of exercise—a potential win for both the efficiency junkies and those with limited mobility. (Just check with your doctor before doing this, especially if you have a medical condition.)

Are You 'Mirin' My Myelin?

Prolactin is a hormone with a wide array of roles that's present in both men and women, but it is perhaps best known for its role in initiating lactation in soon-to-be mothers. It also may have a very interesting influence on the brain: prolactin has been shown to rebuild myelin, the protective sheath that insulates neurons and makes your brain work faster.[14] Women who are pregnant experience a surge in prolactin, and those with MS, an autoimmune disease where myelin is attacked, commonly go into remission at this time.

Not to worry, though—pregnancy is not the only way to boost prolactin. Hyperthermic conditioning has also been shown to boost prolactin dramatically. One study demonstrated that men who stayed in a sauna that was heated to 80°C (176°F) experienced a tenfold increase in prolactin. In a separate study, women who were habitual sauna users who spent twenty minutes in a dry sauna had a 510 percent increase in prolactin immediately following the session.[15]

Can boosting prolactin with saunas be used to treat MS? Great caution should be exercised if the disease has already developed, as temperature-sensitive patients with MS have been shown to exhibit a temporary decline in cognitive function following a sauna. In terms of prevention, sauna use for MS is uncharted territory—but based on the above, its utility is certainly plausible.

ARE YOU SUFFERING FROM CHRONIC CLIMATE CONTROL?

Primates and early humans experienced physiological stressors including changes in temperature for millions of years. Today, however, our lack of such "thermal exercise" may be to the detriment of our health and brain function. But how extreme do temperatures need to be to elicit a positive response from our bodies? Not very, it turns out.

Exposure to even mild ambient cold induces something called *non-shivering thermogenesis*, which is when your body warms itself up to protect against heat loss. Your body does this by ramping up calorie burning in the power-generating mitochondria of brown fat. This is a type of fat that we actually want more of because it promotes better metabolic health. Brown fat is so keen on burning off calories that non-shivering thermogenesis can account for up to 40 percent of your metabolic rate, making it a powerful form of exercise you can do without even moving!

In one shining example of the hormonal benefits of cold exposure, people with type 2 diabetes were told to endure six-hour-per-day exposure to mild cold (60°F). After a mere ten days, they improved their insulin sensitivity by a whopping 40 percent.[16] You may recall from chapter 4 that insulin sensitivity is highly correlated to better brain health and brainpower. Other studies have suggested that thermogenesis (calorie burning in exchange for warmth) and metabolic benefits can occur at an even milder temperature—66°F.

If the thought of being even remotely chilly makes you reach for the nearest blanket, take comfort: the more we expose ourselves to cooler temperatures, the more health benefits we stand to gain. And these benefits increase even as we mentally adapt to the cooler temperatures. So, next time you're standing by your thermostat wondering what to program, keep in mind that chronic climate comfort may be in the same league as sugar when it comes to metabolic mayhem.

Intermittent Fasting

Intermittent fasting is quickly becoming known as one of the best ways to enhance your vitality and vigor. In chapter 6, I discussed how intermittent fasting can stoke the ketogenic fire (your brain's preferred fuel) by reducing insulin. But as a hormetic stressor, fasting is also able to turn on many of the same repair genes that we've already discussed, increasing antioxidant coverage and BDNF production.

It is thought that the body takes these periods of rest from food as an opportunity to clean house, recycle damaged proteins, and kill off immune cells that have become dysfunctional. In antiquity, fasting periods were baked into the cake, so to speak, because food was simply not in plentiful supply year-round. We'll be the first to admit that it's a lot easier to "not eat" when there's no food in sight than to build fasting periods into our busy lives, but as detailed below, we think it's well worth the extra effort.

Whether by time-restricted feeding or periodic low-calorie diets (more on these below), the benefits of fasting are numerous:

- **Improved decision making.**[17] This makes sense from an evolutionary standpoint: what would happen to our survival chances if we got *dumber* the minute food wasn't around? Our species probably wouldn't have lasted very long!
- **Improved insulin sensitivity.** Fasting can improve markers of metabolic health including our ability to effectively use glucose—and fat—as fuel.
- **Enhanced fat loss.** In the morning, cortisol is naturally elevated, allowing for the mobilization of stored fatty acids and sugars that our organs can use for fuel. By fasting, we enable cortisol to do what it does best.
- **Activation of survival genes involved in antioxidant protection and repair.** Intermittent fasting is one of the best ways to jumpstart the Nrf2 pathway, a genetic master switch that increases antioxidant coverage.
- **Activation of autophagy.** Autophagy is the body's waste disposal system, whereby cellular junk (including damaged cells that could lead to cancer) get cleaned up.[18] Much of this debris is pro-inflammatory, and stoking this cleanup process has been associated with dramatically longer life spans and health spans in animals.
- **Improved hormone profile.** Fasting is one of the best ways to boost growth hormone, which is neuroprotective and helps preserve lean muscle tissue.
- **Increased BDNF and neuroplasticity.** Fasting is a potent BDNF booster, which promotes neuroplasticity at any age.

Neuroplasticity is the ability to grow new brain cells and preserve the ones you have, and it even helps to improve mood.

- **Increased cholesterol recycling.** Soon after beginning a fast, the breakdown of excess cholesterol into beneficial bile acids begins.[19]
- **Reduced inflammation and enhanced resistance to oxidative stress.**[20] Studies in humans during the month-long religious holiday Ramadan, which involves daily fasting, have shown that markers of inflammation are dramatically reduced during this period.
- **Enhanced synaptic protection.** New research suggests that fasting might help reduce synaptic activity by preventing excessive neurotransmitter release.[21]

The most popular intermittent fasting protocol is the 16:8 fast protocol, which is a time-restricted feeding diet. This would entail fasting for sixteen hours, while eating unrestrictedly during the eight- (or ten-) hour "feeding" window. This window can be adjusted to whichever hours work best for you,† and women may get just as many benefits from a shorter fast. (As previously discussed, women's hormone systems may be more sensitive to signals of food scarcity. This is just a theory, but women do seem to react differently than men to longer fasts.)

Remember not to deprive yourself during the feeding window. This is when one would consume all of the healthy fats, protein, and fibrous veggies the brain and body need each day. Malnourishment is definitely *not* the end goal! The goal is merely to regain the critical balance between anabolic (storing) and catabolic (breaking down)

† While we surmise that there may be beneficial hormonal effects to keeping the feeding window later in the day rather than eating first thing in the morning (to allow cortisol to do its job of liberating stored fatty acids for use as fuel), you also want to ensure that you leave enough time to digest (two to three hours) before going to sleep. Eating immediately before bedtime can disrupt sleep as well as the brain's maintenance processes.

states. During the fasting window, you may drink as much water as you want, along with tea or black coffee, neither of which contain any calories.

Another protocol is alternate-day fasting, a method studied by University of Illinois researcher Krista Varady. This involves a very small feeding window (between noon and 2 p.m., for example) every other day. This allows for one large meal during fasting days and unrestricted eating on feeding days.[22] There are other protocols that are effective as well, such as back-to-back very-low-calorie days (the *fasting-mimicking diet*, termed by researcher Valter Longo). And, for some people, a full twenty-four- to thirty-six-hour fast every couple of months may be just what it takes to feel a sense of biological "spring cleaning."

Though the intermittent fasting methods are different, the mechanisms are similar, and ultimately it comes down to personal preference. Don't be afraid to play around, and note that many find not eating for a few extra hours per day easier than trying to count calories (including the authors).

"Stressful" Foods

Sola dosis facit venenum. (All things are poison and nothing is without poison; only the dose makes a thing not a poison.)

—PARACELSUS

Stressful foods? I know, that doesn't sound very pleasant. But many of the most valuable foods you already eat every single day confer their benefits by being stressful at the *cellular* level.

Like any organism, plants don't want to be eaten. They're at a bit of a disadvantage, however, as they can't run from predators or fight them off with teeth or weapons. Instead, they turn to chemistry to defend themselves against threats by developing compounds that are toxic to insects, fungi, and bacteria. Many of these natural plant defense chemicals you may already be familiar with: *oleocanthal* from olive oil, *resveratrol* from red wine grapes, and even *curcumin* from turmeric. But in fact, there are thousands of these chemicals that we regularly consume with a vegetable-rich diet, and we're just beginning to understand the impact they have on us—most of them have yet to be named!

Among these chemicals are polyphenols, a large family of plant-based nutrients that are well known to benefit our health. Recent research has highlighted polyphenols as being broadly anti-inflammatory, protecting against age-related inflammation and chronic diseases such as cancer, heart disease, and dementia. Although the exact mechanisms of action behind polyphenol consumption have been somewhat elusive, hormesis has surfaced as one possible explanation.

Here are a few of the common polyphenols, divided by category:

Polyphenol Food Sources	
Catechins	Green and white tea, grapes, cocoa, berries
Flavanones	Oranges, grapefruits, lemons
Flavanols	Cocoa, green vegetables, onions, berries
Anthocyanins	Berries, red grapes, red onions
Resveratrol	Red wine, grape skins, pistachios, peanuts
Curcumin	Turmeric
Oleocanthal	Extra-virgin olive oil

These compounds exert their benefit on us in part by creating a small amount of stress at the cellular level. When we consume polyphenols, our cells respond defensively by flipping the switch on gene activity that boosts antioxidant production. In fact, the antioxidants stimulated by polyphenols outshine the free radical–scavenging effect of more commonly known antioxidants like vitamins E and C. Those antioxidants work "one-to-one"—meaning one molecule of vitamin C disarms one free radical. But the antioxidants that polyphenols lead our bodies to create, such as the free radical–fighting warrior glutathione, can disarm countless free radicals.[23] In this way, eating polyphenol-rich foods is like giving our cells a workout, challenging them to detoxify, adapt, and grow more resilient to stress. (You can further support the production of glutathione—dubbed "the mother of all antioxidants"—by consuming more sulfur-rich foods, including broccoli, garlic, onions, leeks, eggs, spinach, kale, grass-fed beef, fish, and nuts.)[24]

Each polyphenol may have its own unique benefit, but science has revealed some to be particularly beneficial. Oleocanthal in extra-virgin olive oil, for example, has been shown to help the brain clear itself of plaque, stoking the self-cleaning process of autophagy, described earlier. Another phenolic compound called *apigenin*, abundant in parsley, sage, rosemary, and thyme, has been found to promote neurogenesis and strengthen synaptic connections. Perhaps Simon and Garfunkel were inspired by apigenin when they wrote their hit song!

Here are some other well-known polyphenols and their proposed benefits:

Phenol	Found In	Benefits
Resveratrol	Red wine, dark chocolate, pistachios	Improves brain glucose metabolism, cognitive function
Quercetin	Onions	Strengthens gut barrier integrity and reduces permeability
Anthocyanins	Blueberries	Reduces cognitive aging and Alzheimer's risk
Fisetin	Strawberries, cucumbers	Reduces brain inflammation, protects from cognitive decline

ONE MORE REASON TO GO ORGANIC

It's well established that by opting for organic produce, you are avoiding the exposure to synthetic herbicides and pesticides that can disrupt neurotransmitter function and increase your risk for certain neurodegenerative diseases.[25] Here's another reason to look for the organic label: the use of synthetic pesticides and herbicides on produce can dramatically handicap the plants' creation of their own defense mechanisms—the very polyphenols we want.[26] Many studies that compare vitamin content of conventionally grown produce with that of organically grown produce overlook this point. The most health-promoting nutrients in these plants are not always vitamins but the natural defense compounds that, when consumed, stoke genetic restoration pathways in us.

Another well-known class of plant-based defense chemicals are glucosinolates. Cruciferous vegetables like broccoli, cabbage, and kale are rich in these compounds, with young broccoli sprouts taking the crown for the top known source, containing twenty to one hundred times the amount of that of full-grown broccoli heads. When any of these plants are chewed, these compounds combine with an enzyme, also in the plant, that creates a new compound in your mouth—*sulforaphane.*

Thank Darwin you're not an insect, as sulforaphane would be toxic to you if you were! In humans, however, sulforaphane is an anticancer agent, and it also activates the important detoxification pathway *Nrf2*, which dramatically increases glutathione production.[27] Animal studies have repeatedly shown that sulforaphane directly negates inflammation in the brain, even when challenged with highly inflammatory toxins.[28] For this reason, sulforaphane has been studied as a potential therapeutic and preventative agent in

Parkinson's, Alzheimer's, traumatic brain injury, schizophrenia, and even depression—all conditions demonstrated to involve excessive oxidation and inflammation in the brain. One fascinating trial in young people even found that sulforaphane (extracted from broccoli sprouts) significantly improved symptoms of moderate to severe autism. The reduction in symptoms abated after treatment ended.[29]

CRUCIFEROUS VEGETABLES AND YOUR THYROID

Raw cruciferous vegetables like broccoli, cauliflower, kale, bok choy, and cabbage have caught a bad rap, mostly because of compounds within them that are believed to disrupt thyroid function. The compounds in question are the glucosinolates, which are the ones that, when chewed raw, create beneficial sulforaphane.

The issue is that glucosinolates temporarily inhibit uptake of iodine into the thyroid, which is not good because iodine is a required element for thyroid hormone production. In the fifties, when iodine deficiency was widespread, eating otherwise healthy cruciferous vegetables led many to hypothyroidism, and the government thus mandated that all table salt be iodized. Problem solved, right? At the time, yes. Today, however, health-conscious eaters are moving away from iodized salt to non-iodized alternatives like sea salt, and ironically we are again at risk for iodine deficiency. To combat this, it is important to consume sea vegetables (dried nori or kelp noodles are top sources) and other foods rich in iodine such as scallops, salmon, eggs, and turkey. Three ounces of shrimp or baked turkey breast each provide 34 micrograms of essential iodine. That's about 23 percent of the recommended daily intake (RDI). One quarter ounce of seaweed, for comparison, provides 4,500 micrograms of iodine—3,000 percent of the RDI.

In the absence of an iodine deficiency, cruciferous vegetables are perfectly safe to eat raw. The key to remember here is that many of these compounds adhere to a common biological theme: just because you need some doesn't make more *better.* Consume a liberal amount of raw crucifers—just don't go overboard.

So, there you have it—the confidence to know that the right kinds of stress can actually be your friend. These positive stressors provide the key to becoming more robust in brain and body. Remember that in all cases, you should listen to your body, and know that these stresses are not without risk. But if you start slow, coaxing your body to greater resilience, you will get to know the magnitude of your own magnificence in no time.

FIELD NOTES

- Aerobic forms of exercise should be "low and slow" to promote neurogenesis while avoiding the cortisol spike of "chronic cardio."
- Anaerobic forms of exercise should be "hard and fast" to promote metabolic adaptation in your muscles and brain.
- Both forms of exercise are critical!
- Sauna use can be an incredible adjunct to exercise or as a standalone brain booster.
- Fasting helps the body regain its anabolic/catabolic balance, turning on repair genes, burning up stored fuels, and reducing oxidative stress.
- Eat your veggies and low-sugar fruits—they're rich in polyphenols and other compounds that cause your cells to detoxify in powerful ways.

GENIUS FOOD #10

ALMONDS

Aside from being a convenient snack, almonds are a potent brain food for three reasons. First, the skins of almonds have been shown to provide a prebiotic effect, which as you may recall is important for nurturing the mass of bacteria in your large intestine. Researchers fed people almond skins or whole almonds and saw that they both increased populations of beneficial species while reducing pathogenic ones. Second, almonds are a rich source of polyphenols—plant defense compounds that provide an antioxidant effect to both you and your gut bacteria.[1] Last, almonds are a powerful source of fat-soluble antioxidant vitamin E. Vitamin E protects synaptic membranes from oxidation, thus supporting neuroplasticity.[2] Scientists have noted a link between decreasing serum levels of vitamin E and poorer memory performance in older individuals.[3] A 2013 trial, published in the *Journal of the American Medical Association*, even found that high doses of vitamin E led to significantly slower decline in patients with Alzheimer's disease (worth up to six months of bought time).

Almonds do contain substantial amounts of polyunsaturated fat, which as you recall is a fat that is easily oxidized. This is why I prefer to consume almonds, and all nuts, raw. However, for those who prefer roasted nuts, it may provide comfort to know that the fat in almonds remains relatively protected through the roasting process, a sign that the nuts also contain a high amount of antioxidants.[4] Just be sure to go for *dry* roasted nuts, as "roasted" almost always means that they've actually been deep-fried in poor-quality vegetable oil!

How to use: Eat raw as a snack, combine with some dark chocolate and berries for a nice "trail mix," or throw in a salad. Just be mindful that because of their fat content, nuts contain a lot of calories, which can add up fast. Try to stick to a handful or two a day, tops.

Pro tip: All nuts are healthy. While almonds are a great go-to choice, macadamias, Brazil nuts, and pistachios are equally excellent options. Pistachios contain more lutein and zeaxanthin (two carotenoids that can boost brain speed) than any other nut. They also contain resveratrol, a powerful antioxidant that has been shown to protect and enhance memory function.[5]

CHAPTER 11

THE GENIUS PLAN

In this chapter, we're going to put all the pieces from previous chapters together to present the Genius Plan, which will break down the essentials of eating for ultimate cognition nutrition. We'll also discuss various tweaks that you can make to tailor the Genius Plan to your own unique biology and specific cognitive and body goals.

The crux of eating for an optimally performing brain is consuming a diet that is high in nutrient-dense foods (such as eggs, avocados, dark leafy greens, and nuts) and devoid of the foods that cause hormonal dysregulation, oxidative stress, and inflammation (such as processed oils and grain products). Here are some of the things that will immediately begin to happen as you bid adieu to dense, processed carbohydrates and processed oils:

- **You'll lose weight.** Because you will be stimulating insulin to a far lesser degree, it will give your metabolism the chance to liquidate your stored fat and use it for fuel. Recall that insulin is an anabolic (growth) hormone that acts like a one-way valve to your body's fat cells, and lowered insulin levels are a prerequisite to fat burning.
- **Energy and stamina will increase.** Those on high-carbohydrate diets often experience a mental boost when consuming sugar. Does that mean sugar is a performance-boosting agent? No! It is merely treating symptoms of withdrawal. Stepping out of the cycle of carbohydrate addiction is thus the best way to achieve *sustained* high performance.

- **You will be minimizing your risk for prediabetes/metabolic syndrome and, ultimately, type 2 diabetes.**[1] The reduced demand on your pancreas will help promote optimal sensitivity to insulin.
- **If you are already prediabetic or have type 2 diabetes, reducing carbohydrates can help reverse insulin resistance.** Insulin resistance has been associated with greater plaque buildup in the brain and worse cognitive function when compared to metabolically healthy controls. Studies that have pitted the typically prescribed "anti-diabetes" diet (which includes pasta and low-fat tortillas) against a diet devoid of grains (which focuses instead on vegetables and healthy fats) have shown that the grain-free diets improve health outcomes to a greater degree.
- **You will create fewer advanced glycation end-products.** AGEs are *gerontotoxins* that accelerate aging. If the prospect of protecting your eyes, kidneys, brain, liver, and heart doesn't sell you on this point, perhaps the reduction in skin wrinkling and sagging will!
- **You will reduce inflammation throughout your body and as a result may diminish symptoms of conditions caused by inflammation.** Inflammation is a common denominator in many neurodegenerative disorders, including Alzheimer's, Parkinson's, and ALS. It is a major driver of aging, working at the genetic level to make you look, feel, and actually *be* older than you are.
- **You will feel happier and more social.** Inflammation creates "sickness behaviors," designed to prevent further damage, promote healing, and remove you from social settings. This may manifest as reduced cognitive performance, depression, lethargy, inability to focus, and anxiety.
- **Hunger will be a thing of the past.** Although some people adapted to high-carbohydrate diets might experience head-

aches at first, they will soon pass. When all you do is feed your brain glucose, it screams "Feed me!" when it runs out. On the other hand, your body has a virtually unlimited ability to store fat—let it be burned!

- **You will have more room for vegetables on your plate.** Consumption of vegetables and the nutrients they contain is directly associated with brains that work faster and have less risk for developing dementia.

Clear Out Your Kitchen

Cue the opening chords of "Eye of the Tiger." (Prefer "The Final Countdown"? That works too.) You're about to take an inventory of your kitchen and clear out the foods that are no longer serving you. Grab a garbage bag and get ready to load it up—this'll be fun! Start by removing the following:

- **All forms of refined, processed carbohydrates:** This includes products made with corn (and corn syrup), potato flour, and rice flour. These often take the form of chips, crackers, cookies, cereals, oatmeal, pastries, muffins, pizza dough, doughnuts, granola bars, cakes, sugary snacks, candy, energy bars, ice cream and frozen yogurt, jams/jellies/preserves, gravies, ketchup, honey mustard, commercial salad dressings, pancake flours and mixes, processed cheese spreads, juices, dried fruit, sports drinks, soft drinks/soda, fried foods, and frozen packaged foods.
- **All sources of wheat and gluten:** Bread, pasta, rolls, cereals, baked goods, noodles, soy sauce, and anything with wheat flour, enriched wheat flour, whole-wheat flour, or multigrain flour in its ingredients list. Most oatmeal contains gluten unless it explicitly says "gluten-free" on the label.

- **Sources of industrial-grade emulsifiers:** Anything with polysorbate 80 or carboxymethylcellulose in the ingredients list. Common offenders include ice cream, coffee creamers, nut milks, and salad dressings.
- **Industrial and processed meats and cheeses:** Grain-fed red meat, feedlot chicken, processed cheeses.
- **All concentrated sweeteners:** Honey, maple syrup, corn syrup, agave syrup or nectar, simple syrup, or sugar, both brown and white. (Don't worry, I'll offer some safe noncaloric sweetener options momentarily.)
- **Commercial cooking oils:** Margarines, buttery spreads, cooking sprays, and oils like canola, soybean (sometimes labeled "vegetable oil"), cottonseed, safflower, grapeseed, rice bran, wheat germ, and corn. Even if they're organic, toss them. Remember that these oils are often included in various sauces, mayonnaise, and salad dressings and serve no purpose to you other than to supply you with damaged, oxidative omega-6 and omega-3 fats. Get your omegas from whole-food sources instead.
- **Nonorganic, nonfermented soy products:** Tofu.
- **Synthetic sweeteners:** Aspartame, saccharin, sucralose, acesulfame-K (also known as acesulfame potassium).
- **Beverages:** Fruit juice, sodas (diet and regular), commercial fruit smoothies.

Always Foods: Stock Up

These are the foods that can be consumed liberally through all phases of the plan. Calorie counting is usually unnecessary; however, if your goals include losing weight, consume fewer concentrated fats (oils, butter, and so forth). If you are looking to maintain or gain weight, more fats may be included. Keep in mind: with the exception of extra-virgin olive oil, we are not necessarily in favor of a high *added*-fat diet, as pure oils are not very nutrient dense.

- **Oils and fats:** Extra-virgin olive oil, grass-fed tallow and organic or grass-fed butter and ghee, avocado oil, coconut oil.
- **Protein:** Grass-fed beef, free-range poultry, pasture-raised pork, lamb, bison, and elk, whole eggs (revisit the healthy egg matrix on page 144), wild salmon, sardines, anchovies, shellfish and mollusks (shrimp, crab, lobster, mussels, clams, oysters), low-sugar beef or salmon jerky.
- **Nuts and seeds:** Almonds and almond butter, Brazil nuts, cashews, macadamias, pistachios, pecans, walnuts, flaxseeds, sunflower seeds, pumpkin seeds, sesame seeds, chia seeds.
- **Vegetables:** Mixed greens, kale, spinach, collard greens, mustard greens, broccoli, chard, cabbage, onions, mushrooms, cauliflower, Brussels sprouts, sauerkraut, kimchi, pickles, artichokes, alfalfa sprouts, green beans, celery, bok choy, watercress, asparagus, garlic, leeks, fennel, shallots, scallions, ginger, jicama, parsley, water chestnuts, nori, kelp, dulse seaweed.
- **Nonstarchy root vegetables:** Beets, carrots, radishes, turnips, parsnips.
- **Low-sugar fruits:** Avocados, coconut, olives, blueberries, blackberries, raspberries, grapefruits, kiwis, bell peppers, cucumbers, tomatoes, zucchini, squash, pumpkin, eggplant, lemons, limes, cacao nibs, okra.
- **Herbs, seasonings, and condiments:** Parsley, rosemary, thyme, cilantro, sage, turmeric, cinnamon, cumin, allspice, cardamom, ginger, cayenne, coriander, oregano, fenugreek, paprika, salt, black pepper, vinegar (apple cider, white, balsamic), mustard, horseradish, tapenade, salsa, nutritional yeast.
- **Fermented, organic soy:** Natto, miso, tempeh, organic gluten-free tamari sauce.

- **Dark chocolate:** At least 80% cocoa content (ideally 85% or higher).
- **Beverages:** Filtered water, coffee, tea, unsweetened almond milk, unsweetened flax milk, unsweetened coconut milk, unsweetened cashew milk.

Sometimes Foods: Eat in Moderation

These foods should be included in moderation, consumed later in the day and only after the initial two-week ultra-low-carb break-in. Moderation means at most a few (three to four) servings a week. Again, choose organic if possible.

- **Starchy root vegetables:** White potatoes, sweet potatoes.
- **Non-gluten-containing unprocessed grains:** Buckwheat, rice (brown, white, wild), millet, quinoa, sorghum, teff, gluten-free oatmeal, non-GMO corn or popcorn. Oats do not naturally contain gluten but are frequently contaminated with gluten as they are processed in facilities that also handle wheat. Therefore, look for oats that explicitly indicate on the package that they are gluten-free.
- **Dairy:** Grass-fed, full-fat, and antibiotic- and hormone-free yogurt, heavy cream, and hard cheeses are acceptable.
- **Whole, sweet fruit:** While low-sugar fruits are *always* the best choice, apples, apricots, mangos, melons, pineapple, pomegranates, and bananas provide various nutrients and different types of fiber. Be extra-cautious with dried fruit, which has the water removed and sugar concentrated, making it easy to overdo it. These are all best consumed after a workout.
- **Legumes:** Beans, lentils, peas, chickpeas, hummus, peanuts.
- **Sweeteners:** Stevia, non-GMO sugar alcohols (erythritol is best to use, followed by xylitol, which is naturally harvested from birch trees), monk fruit (*luo han guo*).

It is essential that any corn and soy products, if consumed at all, are organic and non-GMO, as these two commodities tend to be the most manipulated to withstand the heavy use of pesticides and herbicides.

Remember that once the brain has become fat adapted, a higher-carb meal here and there (particularly when timed around exercise) will not throw you off. At that point, consumption of the foods on the above list can be increased, but the goal should generally be less than 75 grams of net carbohydrates (total carbohydrate content minus grams of fiber) per day.

FAQ: I'm extremely active—doesn't that mean I can eat more carbs?

A: Yes, engaging in vigorous exercise allows you more leeway—see our Custom Carb Pyramid (page 318) for exact numbers. Most people are not very active, however, and even those who think they are aren't compared to our ancestors.

Meal Planning

Breakfast

There is *no biological need* to eat first thing in the morning. Breakfast in its most common forms only helps you store fat.[2] The best breakfast will often be a glass of water, black coffee, or unsweetened tea. If you choose to eat breakfast, ensure that it is primarily protein, fat, and fiber. (Example: my "Cheesy" Scrambled Eggs on page 323.)

Lunch

Here are some great lunch options:

- A big salad with grilled chicken (see my Huge Daily "Fatty" Salad rule on page 310).
- A roasted vegetable bowl with pastured pork belly, wild salmon, or grass-fed beef.
- A whole avocado plus a can of wild sardines.

Dinner

Load up on veggies and properly raised sources of protein. Eat to your heart's content! And don't forget to use extra-virgin olive oil liberally as a sauce (you can use a full tablespoon or two per person). Here are a few great dinner examples:

- Roasted Brussels sprouts with extra-virgin olive oil and Grass-fed Picadillo (page 325).
- Sautéed Greens (page 332) with extra-virgin olive oil and salt-and-pepper-seasoned wild salmon.
- Huge "Cheesy" Kale Salad (page 334) with Insanely Crispy Gluten-Free Buffalo Chicken Wings (page 329).

Snacks

- Blueberries
- Jicama sticks
- Dark chocolate
- Half an avocado with sea salt
- Nuts and seeds
- Low-sugar beef or salmon jerky
- Celery with raw almond butter
- A can of wild sardines in extra-virgin olive oil (personal favorite!)

- Pastured pork rinds sprinkled generously with nutritional yeast (also great!)

An Example Genius Week

See chapter 12 for many of the recipes detailed below.

MONDAY

Morning: water, black coffee or tea

First meal: 2 or 3 eggs, $\frac{1}{2}$ avocado

Snack: $\frac{1}{2}$ avocado sprinkled with sea salt and drizzled with EVOO

Dinner: wild salmon fillet, large fatty salad

TUESDAY

Morning: water, black coffee or tea

First meal: Better Brain Bowl (page 333)

Snack: handful of raw nuts, blueberries, a few squares of dark chocolate

Dinner: grass-fed beef burger, hummus, sautéed greens

WEDNESDAY

Morning: water, black coffee or tea, fasted workout

First meal: large fatty salad, large sweet potato

Snack: can of sardines or wild salmon

Dinner: Banging Liver (page 328), roasted Brussels sprouts

THURSDAY

Morning: water, black coffee or tea

First meal: over-easy eggs with kimchi and EVOO

Snack: celery with raw almond butter and cacao nibs

Dinner: Jamaican Me Smarter (page 324), sautéed greens

FRIDAY

Morning: water, black coffee or tea, fasted workout

First meal: "Cheesy" Scrambled Eggs (page 323), large sweet potato, 1/2 avocado

Snack: low-sugar beef jerky, bottle of kombucha

Dinner: Insanely Crispy Gluten-Free Buffalo Chicken Wings (page 329), sautéed greens

SATURDAY

Morning: water, black coffee or tea

First meal: 3 scrambled eggs with vegetables

Snack: pastured pork rinds with nutritional yeast

Dinner: huge fatty salad, can of sardines

SUNDAY

Morning: water, black coffee or tea

First meal: poached eggs over sautéed greens, EVOO

Snack: whole avocado with sea salt, handful of nuts

Dinner: skip

A Note on Nut Milks

While unsweetened nut milks are approved under the Genius Plan, make sure yours are free of the very commonly used emulsifiers polysorbate 80 and carboxymethylcellulose. These chemicals, used to create a creamy mouthfeel in processed foods, have been shown in animal models to cause inflammation and metabolic dysfunction through the gut, thus posing a potential threat to your brain. You can read more about the deleterious effects of emulsifiers on page 198.

Also keep in mind that an eight-ounce cup of almond milk pales in comparison, nutritionally, to even a tiny handful of actual almonds, while being about ten times more expensive—one gallon of almond milk contains approximately thirty-nine cents' worth of almonds!

Opt for Organic

Choose organic foods whenever possible; however, if cost presents an issue, simply look up the Environmental Working Group's most current "Dirty Dozen" and "Clean Fifteen" (the EWG puts out a new list every year), which group conventionally grown produce in terms of least ("clean") and highest ("dirty") pesticide content. Here is an abbreviated list of brain-optimizing foods as of this writing:

Dirty—Should always be organic	Clean—Needn't be organic
Kale	Asparagus
Collard greens	Avocados
Spinach	Cabbage
Strawberries	Cauliflower
Cucumbers	Onions
Bell peppers	Eggplant
Cherry tomatoes	

Divide and Conquer . . . Your Plate

In regard to the ratio of animal protein to vegetable intake, you will be consuming mostly vegetables by volume and mostly fat by calories. This is because vegetables are satiating but do not provide very many calories. Fats will make up the majority of calories consumed throughout the day, but when looking at a plate, most of the real estate will be given to colorful, fibrous vegetables. Eating mostly vegetables also helps neutralize oxidative free radicals that are generated during the cooking process (in meats, for example) before they're ever absorbed into the bloodstream.

Abide by the "One Bad Day" Rule

Animals raised on local farms and allowed to eat their preferred diets are happier *and* healthier. Many local farmers take great care in treating their animals well, priding themselves on the fact that their livestock have only "one bad day." This is a stark contrast to the way the vast majority of livestock are raised today, forced to live out lives of misery in cramped cages, fed diets that make them sick, and minimally exposed to the outside or even one another. While eating animals may have been an essential part of our evolution, to be *humane* is an essential part of being human—and conveniently, the humane choice is healthier for you and the environment.

I suggest abiding by the following rule: only consume meat where you can be sure that the animal had just "one bad day."

The Huge Daily "Fatty" Salad

One of the best strategies to cover your food bases is to eat a huge salad every single day, and load it up with healthy fats and protein. Though eating salads for better health may seem pretty intuitive, by creating a rule for yourself to incorporate one large salad every single day, you are ensuring that you enrich your brain with a diverse array of plant nutrients and fiber. Plus, there is simply no better vehicle for extra-virgin olive oil than a salad!

Whether for lunch or dinner, every salad is a new opportunity to feed your brain (and your gut microbes). Make sure that you own a very large bowl (the bigger the better, and I like glass so that I can see all of the colors I'm eating), and go to town. For the base, opt for nutrient density—skip pale white iceberg lettuce, which is nutritionally poor and mostly water, and reach for the darker leaves instead. Spinach and kale are great options. Here are two ideas—feel free to improvise on them:

- Kale, cucumber, thinly sliced jalapeño chiles, raw broccoli, sunflower seeds, avocado, grilled chicken, extra-virgin olive oil, balsamic vinegar, salt, pepper, lemon
- Spinach, arugula, tomatoes, bell peppers, chia seeds, avocado, grilled shrimp, extra-virgin olive oil, balsamic vinegar, salt, pepper, minced raw garlic, lemon

The beauty of crafting salads is that there are no rules! Throw together as many veggies as you can and drench them with olive oil, which will increase the absorption of their many nutrients (including carotenoids, which can boost brain-processing speed). The key is to aim for one huge salad daily, and there is plenty of room for healthy variety.

What's the Deal with Dairy?

Seventy-five percent of the global adult population is believed to be lactose intolerant, and Harvard's School of Public Health has recently banned dairy from its "Healthy Eating Plate." So what gives?

Milk protein is on par with white bread in terms of insulin stimulation, and from an evolutionary sense this is likely as a means of helping a newborn pack on weight. But bovine milk proteins specifically metabolize into morphine-like compounds called *casomorphins*, which seem to have an inflammatory effect on the gut. They've also been shown to interact with neurotransmitters and have been linked with headaches, delayed psychomotor development, autism, and type 1 diabetes.[3]

The direct effect that this has on the average brain has yet to be borne out in the literature, but there is one other line of research that is worth mentioning. Milk lowers a compound in the body called *urate*. Very high levels of urate can cause gout, but at normal levels, the chemical seems to be a powerful antioxidant for the brain, and particularly protective against Parkinson's disease. Both milk consumption and reduced urate levels have been linked with higher risk for developing Parkinson's disease, and studies are under way now to see if raising urate can slow Parkinson's progression.

For these reasons, I don't recommend dairy other than butter and ghee. But if you aren't sensitive to it and choose to enjoy it occasionally, stick with full-fat varieties.

Steer Clear of Fake Gluten-Free Foods

Substituting gluten-containing foods with highly processed gluten-free doppelgängers (such as most gluten-free cookies and bread products) is assuredly not the way to go—these foods, often made with highly processed grain flours and refined sugar, can be profound blood sugar boosters, negating nearly if not all of the benefits of going gluten-free for the nonceliac population. In addition, they commonly contain easily oxidized polyunsaturated fats, which may contribute to free radical cascades in your arteries. Always stick with the foods that never had gluten in them to begin with—not industrially manipulated food-like approximations of the real thing.

And What about Alcohol?

On the one hand, research has shown that moderate drinkers of alcohol (up to two glasses a day for men, one for women) tend to have better health. On the other hand, ethanol (which is what gives us the "buzz") is a neurotoxin, and when looking specifically at brain health, the research is somewhat less peachy: a thirty-year study found that even moderate alcohol drinkers (who consumed five to seven drinks per week) had triple the risk of hippocampal shrinkage than abstainers.[4]

The psychological benefits of moderate alcohol consumption as a social lubricant and de-stressor are not trivial. In an ideal world, we would all have healthy stress-coping mechanisms and drink minimally, having one to two drinks per week at most—but we also aren't living stress-free existences, frolicking through the woods and picking berries all day either. While I recommend abstaining from alcohol, if you choose to imbibe, here are some tips to make drinking as brain-healthy an experience as possible:

- **Always ensure that you go to sleep *sober*.** Alcohol markedly diminishes sleep quality and affects various hormones that are released during sleep, notably growth hormone.[5]

- **Follow the "one-for-one" rule.** In between each drink, always consume a glass of water. Alcohol irritates the gut and makes it harder to rehydrate once the damage has been done.
- **Sprinkle a little salt in that water.** Alcohol is a diuretic, which can cause you to excrete electrolytes like sodium. Make sure to replace what is lost with a little salt.
- **Stick to red wine, dry white wine, or spirits.** Stick with the spirit of choice "on the rocks," or with soda water and a lime. Avoid sugary mixers like juice or soda at all costs.
- **Drink on an empty stomach.** This may be a more controversial tip, but drinking on an empty stomach may allow the liver to more efficiently process the alcohol without impeding digestive processes. Alcohol impairs LDL recycling and increases post-meal triglyceride (fat in the blood) spikes. Have the drink before or after dinner, not during—just exert caution because a drink can feel more potent on an empty stomach.
- **Avoid gluten-containing beverages, which may be a one-two punch.** Gluten increases gut permeability, which may compound the same effect from alcohol. Beer drinkers, I'm looking at you.

The Medicine Cabinet

Ensuring healthful products in the bathroom is a means of "dotting your i's and crossing your t's" for long-term health and moment-to-moment wellness and performance. Here are a few changes that will make the most impact.

- **Switch to an aluminum-free deodorant.** Many deodorants contain aluminum, and excessive exposure to aluminum has been strongly linked with increased dementia risk. While research has yet to confirm causality, why take the chance? *Alternative:* Buy aluminum-free deodorant, or make your own with coconut oil (a selective bactericide) and baking soda.

- **Avoid frequent use of nonsteroidal anti-inflammatory drugs (NSAIDs) for pain relief.** Regular use of NSAIDs like ibuprofen and naproxen has been linked recently to an increased risk of cardiac events. While these drugs are commonly used to treat minor aches and pains, they "attack" cell mitochondria, reducing their ability to produce energy and increasing the production of reactive oxygen species (or free radicals). This was shown in heart cells, but these drugs can easily cross the blood-brain barrier. *Alternative:* Try curcumin, an anti-inflammatory that has been found to reduce pain, instead. Omega-3 EPA may help as well, as it is a potent anti-inflammatory.
- **Avoid chronic use of acetaminophen.** Acetaminophen, a common over-the-counter pain reliever, can diminish the body's supplies of glutathione, a master brain antioxidant. *Alternative:* Curcumin or EPA.
- **Stop using anticholinergic drugs (described in chapter 8).** These drugs are commonly used to treat allergy symptoms or as nighttime sleep aids and block the neurotransmitter acetylcholine, which is important for learning and memory. *Alternative:* Consult with your physician if these medications are prescribed.
- **Ditch acid blockers, especially proton pump inhibitors (PPIs).** These drugs are often taken for acid reflux, but can alter digestion, blocking the absorption of vital nutrients like B_{12}, thereby increasing risk for cognitive dysfunction and dementia. *Alternative:* By reducing carbohydrate intake, you will likely reduce your symptoms of reflux as well as the need for medication.[6]
- **Avoid antibiotics, especially broad-spectrum varieties, unless necessary.** *Alternative:* Ask your doctor for a narrow-spectrum antibiotic.

Days 1 to 14: Clearing the Cache

Now that you've cleaned out your kitchen and medicine cabinet and stocked up on brain-boosting foods, it's time to kick off the first two weeks of the Genius Plan.

In the first week, the focus should be on eliminating the junk foods in your diet and turning instead to cognition-boosting and fat-burning foods. The foods we will remove first are those for which there is absolutely no human biological requirement: that means all processed foods and anything containing refined wheat and grains, seed and grain oils, and added sugar (including beverages!) should be cut. By dumping these foods from our diet, we're skimming off the majority of calories consumed by most people in the Western world. These are calories that come from "ultra-processed" foods, the vilest offenders. These foods digest rapidly, sending blood sugar levels soaring, and promote large spikes of insulin as a counterbalance, creating fatigue from the blood sugar roller coaster—and you should kick them all to the curb for good during week one.

During this first week, we'll also begin an ultra-low-carb phase that will last through week one and week two. That means we will be eliminating all non-gluten-containing grains, legumes, and other sources of concentrated plant sugars, including tubers and sweet fruits. This is important to help reset the body's metabolism to its "factory settings" and convert a body that is used to burning carbohydrates for fuel to one that is fat adapted and thus metabolically flexible. This break-in period will still include all the carbohydrates you need in the form of fibrous veggies and low-fructose fruits. In this ultra-low-carb phase, net carbohydrate intake (which is total carbohydrates minus dietary fiber) can range from 20 to 40 grams per day and should consist mainly of green, nonstarchy vegetables. Initially, the fewer the carbs the better—and don't worry, because in the third week we begin to reintegrate carbohydrates to support our activity levels.

Over the course of these two weeks, as your omega-3–to–omega-6 ratio begins to align with what is biologically appropriate, you will begin to experience greater mental stamina and focus and improved mood. By the end of week two, you should experience not only improved digestion from the increased vegetable fiber but deeper sleep. Recent studies have shown that fiber consumption can increase sleep quality, particularly time spent in slow-wave sleep.[7] This is when growth hormone secretion is at its peak and the brain is cleaning itself of waste products accumulated during the day. Upon waking you should feel more rested and experience greater mental edge.

AVOIDING THE LOW-CARB FLU

Some people who are transitioning to a low-carb diet for the first time experience withdrawal symptoms similar to a drug addict quitting cold turkey. In the past, you might have "self-medicated" with carbs the moment blood sugar would drop, but this only serves to perpetuate the vicious cycle. Here, the strategic use of coconut or MCT oil may help to wean the brain off of glucose as your fat-burning machinery ramps up. During this initial low-carb phase in the first two weeks, I recommend 1 to 2 tablespoons of coconut or MCT oil per day. Start slow to avoid an upset stomach, which can occur with excessive MCT oil consumption!

Also, lowering insulin (which will occur during this period) will cause your kidneys to excrete sodium, adding to the "flu." Therefore, you may want to increase your salt intake. I detailed this overlooked fact on page 158; in essence, during the first week of carbohydrate restriction, you may require up to an *additional* 2 grams of sodium—about a teaspoon of salt—per day to feel optimal, which can be reduced to 1 gram after the first week (1/2 teaspoon).

During this phase, it is fine to include one to two cups of coffee per day, but any more than that should be curtailed. While coffee has many brain-protective compounds and has been regarded quite favorably by recent research, it is nevertheless a stimulant to the central nervous system and can disrupt the natural balance between your sympathetic (fight-or-flight) and parasympathetic (rest-and-digest) nervous systems. Also, try to avoid consuming any coffee after 2 p.m. so as not to interfere with sleep. Once a month, it may even be beneficial to switch to decaf for a week as a means to reset caffeine tolerance. You likely won't even notice the difference—never underestimate the unconscious power of classical conditioning!

Days 15+: Strategically Reintegrate Carbohydrates

At this point, you've been on an ultra-low-carbohydrate, high-fiber diet for two weeks. You have likely become metabolically adapted to burning fat for fuel. Here is when you can begin adding in higher-carb, lower-fat "refeed" meals a few days per week (see the Custom Carb Pyramid, page 318, for details). Carbs and insulin are not evil—they are just overabundant and misused today. Strategically integrating them will serve two purposes: refilling muscles with stored glycogen (sugar), as well as upregulating hormones that can become lowered by extended low-carb dieting, including leptin, the master metabolic regulator.

HOW (AND WHEN) TO DO A CARB REFEED

Not everybody may need to reintegrate starches after two weeks. If you are overweight or insulin resistant, it may be more important to continue following an extremely carbohydrate-restricted diet (20 to 40 grams of net carbs per day) to drop the excess weight and regain metabolic resilience.

Your goal should be to become insulin sensitive first (i.e., to reduce fasting insulin and glucose) *before* experimenting with higher-carb refeeds.

For a metabolically healthy, fat-adapted person (revisit page 160 to see what being fat adapted means and how it should feel), an occasional post-workout higher-carb, low-fat meal may be beneficial. For example, with high-intensity exercise, post-workout carbs can actually help increase performance. Normally, cells require insulin to wrangle glucose transporters and bring them to the surface of cell membranes, but in the window following strength training, muscles act as a sponge for sugar, pulling glucose out of the blood without the need for insulin. These carbohydrates are less likely to be stored as fat, and reentering fat-burning mode will be much hastier. The resultant increase in muscle mass will increase your overall metabolism and provide an additional buffer for excess calories.

Ripe spotted bananas, berries, white or brown rice, and starchy vegetables and other low-fructose foods are excellent choices for a carb refeed, and anywhere between 75 and 150 grams of net carbs may be consumed to provide an anabolic stimulus without compromising fat adaptation. (This is still *dramatically* less than the standard American carbohydrate intake of 300-plus grams per day.) Individual experimentation may be warranted, but try to keep the consumption of these carbs close to the exercise session to help minimize fat storage. Depending on how advanced your training is, these refeeds can be integrated once to a few times per week.

Note: The science on carb refeeds is far from settled, but it certainly suggests that occasional insulin spikes are not harmful and in fact are important for anabolic stimulus, testosterone and thyroid function, and lean mass preservation. That said, we always want to reduce the net insulin secreted and avoid the frequent hills and valleys that come from multiple carb loads over the course of a day.

Custom Carb Pyramid

Given the wide variability of body types and genetics, use the rough guidelines below to experiment with and determine your optimal carbohydrate intake for metabolic flexibility. **Carb intake (three tiers):**

ULTRA-LOW/KETOGENIC (DAYS 1 TO 14)

- Consume only 20 to 40 grams of carbohydrates per day.
- Stick with this tier for the first ten to fourteen days to deplete glycogen (stored sugar) and fat-adapt the brain.
- If you are eating this way longer-term for weight loss, add a one-time weekly high-carb refeed. This means you can indulge in higher-starch foods (aim to keep the meal low-fat) to replenish muscle energy stores once a week. There's no magic number, but shoot for 100 to 150 grams of carbs with that meal.

LOW(ER) CARB (AFTER 14 DAYS)

- Consume 50 to 75 grams of carbohydrates per day.
- People who are looking for weight maintenance and perform light physical activity should stay at this level.

OPTIONAL: CARB CYCLING

- In this tier, carb intake can be increased after vigorous training (see anaerobic examples in chapter 10 and below).
- Consume 75 to 150 grams of carbohydrates per day.
- This amount is still vastly lower carb than the average American diet. You can take advantage of the boosts that you get

from carbs by mixing low-carb days with higher-carb days to fuel workouts and muscle growth, and to maintain muscle when you're losing body fat.

On heavy workout days, start with 100 to 150 grams post-workout, and have a lower fat intake that day. Glycogen-depleting workouts include multiple heavy sets of compound movements. This means 40 to 70 reps per muscle group, two to three muscle groups per workout, and compound movements including barbell squats, deadlifts, pull-ups, push-ups, bench press, lunges, and dips. Dr. Paul and I recommend working with an experienced trainer if you are lifting for the first time.

- Protein intake:
  - Start at 0.5 grams per pound of body weight. You can increase to 0.8 grams per pound if losing or gaining weight or if doing heavy weight training.
- Meal timing and frequency:
  - Eat fewer carbs pre-workout and more post-workout.
  - Try to concentrate carbs into a single sitting to avoid prolonged insulin spikes.
  - Eat two to four meals per day.
- Fasting:
  - Choose a feeding window (eight hours for men, ten for women, as an example), and consider skipping breakfast.
  - Experiment with different protocols to see what works best (there are a few options detailed in chapters 6 and 10).
  - When fasting, be sure to drink lots of fluids, and add electrolytes like salt.

Example week:

	Sunday	Monday	Tuesday	Wednesday	Thursday	Friday	Saturday
Exercise	Long hike or walk	Resistance training	Yoga, walking	Bike ride	Resistance training	Yoga	Sprints in the park
Carbs	20 to 40 grams (low carb)	150 grams (higher carb)	20 to 40 grams (low carb)	20 to 40 grams (low carb)	150 grams (higher carb)	20 to 40 grams (low carb)	75 grams (low to moderate carb)
Meals	2 meals	3 meals	3 meals	3 meals	3 meals	3 meals	2 meals

Final Notes

As they say in the film industry, "That's a wrap!" I hope that in reading *Genius Foods* you have learned as much as I have in researching, writing, and living the ideas presented here. Collaborating with Dr. Paul was mostly pleasant as well. (I kid—it was great.)

Remember: nutrition is a constantly evolving science—one where there are seldom black-and-white truths. In life, and especially on the Internet, people tend to be religious about their nutrition beliefs. But science is meant to be dispassionate—a method of asking questions and seeking answers, even if those answers are not what you want to hear. I ask that you seek your own truth. Challenge your assumptions regularly, be unafraid of authority, and question everything—*even* what you read in books (including this one).

I'm humbled and honored that you chose to read *Genius Foods* (and I hope you'll consider recommending it to a friend or a loved one—the ultimate form of praise). As fun and fascinating as it was for me to research and write *Genius Foods*, it was driven by my wish that my mom's health could be back the way it was. I wrote this book with the sole intent of helping others to feel better and suffer less. This way, none of it was in vain.

Now, I implore you to take these findings and write your own health story.

CHAPTER 12

RECIPES AND SUPPLEMENTS

Learning how to cook healthy foods that you enjoy is one of the greatest gifts that you can give yourself. It also gives you an excuse to invite friends over and throw dinner parties, which are not only a lot of fun but good for you. In this section, I'm going to share some recipes I've created and some that were contributed by my very talented friends.

Recipes

"Cheesy" Scrambled Eggs

I could eat these every day. Here's one of the best tips to make killer eggs: the lower the heat, and the slower you cook them, the better and creamier they'll be. And always remove them from the heat just prior to your desired level of doneness (as eggs continue to cook for a moment after removing them from the heat).

SERVES 1

What you'll need:

1 tablespoon plus 1 teaspoon avocado oil or extra-virgin olive oil

3 whole pastured or omega-3 eggs, beaten

1½ teaspoons nutritional yeast
3 pinches of salt

What to do:

1. Heat 1 tablespoon of the oil in a large skillet over very low heat. Add the eggs to the pan and slowly scramble using a heatproof spatula. Sift the nutritional yeast over the eggs and stir it in. Add 2 pinches of salt.
2. Remove from the heat just prior to reaching your desired consistency.

How to serve it:

1. Drizzle the remaining 1 teaspoon of oil on top and finish with a pinch of salt. I often serve my eggs with a whole sliced avocado on the side. For variation, throw some diced onions, chopped bell peppers, or sliced mushrooms into the pan and sauté them before adding the eggs.

Jamaican Me Smarter

When I was a kid in New York, one of my absolute favorite after-school snacks was the Jamaican beef patties I'd buy from the local pizza shops. Delicious as they were, they probably were loaded with trans fats and processed oils. Here, I've recreated the seasoning of the beef, and I love to eat it over sautéed vegetables. This is one nourishing dish.

SERVES 2–3

What you'll need:

1 teaspoon ghee
½ yellow onion, chopped

5 garlic cloves, smashed and peeled
1 pound grass-fed ground beef
1 teaspoon salt
1 tablespoon ground cumin
1½ teaspoons ground turmeric
½ teaspoon ground coriander
½ teaspoon ground allspice
½ teaspoon ground cardamom
¼ teaspoon freshly ground black pepper
¼ cup nutritional yeast, optional but recommended

What to do:

1. Heat the ghee in a medium skillet over medium heat. Add the onion and cook for 4 to 5 minutes, until softened. Throw in the smashed garlic and allow to aromatize for 1 minute. Add the ground beef, throw on the salt and all the spices, and cook, stirring often to break it up, until browned, about 10 minutes. Optional: sprinkle on a generous amount of nutritional yeast.

How to serve it:

1. Alongside or over Sautéed Greens (page 332); my favorite is kale.

Grass-Fed Picadillo

I lived in Miami for four years when I went to college and couldn't get enough of Cuban food, especially picadillo. Here's a healthy variation on this traditional dish that I make often.

SERVES 2–3

What you'll need:

1 tablespoon extra-virgin olive oil

1 large yellow onion, finely chopped

4 garlic cloves, smashed and peeled

1 pound grass-fed ground beef

1 teaspoon salt

1½ teaspoons freshly ground black pepper

¼ teaspoon red pepper flakes, optional

⅓ 12-ounce jar no-added-sugar organic tomato sauce (tomato sauce will have some natural sugar from the tomatoes)

½ cup pitted olives, sliced (olives stuffed with pimientos is fine)

What to do:

1. Heat the oil in a large skillet over medium heat. Add the onion and cook for 4 to 5 minutes, until softened. Throw in the smashed garlic and allow to aromatize for 1 minute. Add the ground beef, throw on the salt, pepper, and red pepper flakes, if using, and cook, stirring often to break it up, until browned, about 10 minutes. Add the tomato sauce and olives, bring to a simmer, then reduce the heat to very low and simmer for 10 minutes.

How to serve it:

1. Alongside or over Sautéed Greens (page 332) or "riced" cauliflower (sautéed in garlic, salt, and extra-virgin olive oil).

Pan-Seared Wild Alaskan Salmon with Turmeric, Ginger, and Tahini-Miso

Now that you know that wild salmon is a Genius Food, let me teach you how to turn your average fillet into a superfood and mega-nutrient-rich meal in only a few steps. This recipe was contributed by my good friend and wellness chef Misha Hyman.

SERVES 2–3

What you'll need:

SALMON:

1 pound fresh or frozen wild Alaskan salmon
Salt to taste
Coarsely ground black pepper to taste
Extra-virgin olive oil

TAHINI-MISO:

1/4 cup tahini
1/2 cup brown rice miso
1/4 cup toasted sesame oil
Grated ginger to taste
Grated garlic to taste
Grated fresh turmeric to taste
Fresh lemon juice

GARNISHES:

Handful of finely chopped scallions
1 teaspoon chopped fresh cilantro
Handful of black sesame seeds

What to do:

1. **Get the salmon ready:** Take fresh salmon out of the fridge about an hour before you begin cooking it so it can come to room temperature. This is important because you want your fish to cook evenly. If you are using frozen salmon, defrost it fully and bring to room temperature. Sprinkle the salmon with salt and pepper. Don't be shy.
2. Preheat the oven to 425°F.
3. **Make the tahini-miso:** Combine all the ingredients in a blender and blend until smooth. Set aside while you cook the salmon.
4. **Cook the salmon:** Place the oil in a pan and heat over medium heat. When hot, place the salmon into the pan with the skin facing up. Cook for 3 to 4 minutes, then transfer to the oven and cook for 6 to 8 minutes depending on how well-cooked you like your fish.
5. Immediately brush the salmon with a thin coating of tahini-miso. Garnish the with scallions, cilantro, and black sesame seeds.

How to serve it:

1. This salmon goes wonderfully with asparagus sautéed in grass-fed butter with garlic and turmeric, with fresh spinach tossed in at the end to wilt it, and sprinkled with hemp seeds.

Banging Liver

This is a recipe from my friend Mary Shenouda, aka @PaleoChef on Instagram. I'd never tasted chicken liver before trying Mary's dish, but it made me an instant convert. It is both delicious and a "banging" source of nutrients including choline, vitamin B_{12}, folate, and vitamin A.

SERVES 2–3

What you'll need:

1 pound organic chicken livers, chopped

3/4 teaspoon salt

1/3 cup ghee

6 garlic cloves, crushed and minced

1 large green bell pepper, chopped

1 jalapeño chile, seeded and chopped

1 tablespoon ground cumin

1/2 teaspoon ground cinnamon

1/4 teaspoon ground ginger

1/4 teaspoon ground cloves

1/4 teaspoon ground cardamom

Juice of 1 lime

What to do:

1. Clean and roughly chop the liver. Sprinkle with the salt, give it a toss, and set aside for 2 to 3 minutes.
2. Heat the ghee in a large skillet over medium-high heat, add the liver, and sear until browned on both sides. Add the garlic, bell pepper, and jalapeño and cook until the vegetables are starting to soften, about 5 minutes. Add the cumin, cinnamon, ginger, cloves, and cardamom, reduce the heat to medium-low, cover, and cook for another 5 to 8 minutes. Add the lime juice, scraping up any browned bits from the bottom of the pan and mixing well. Remove from the heat.

How to serve it:

1. Serve with an additional hit of melted ghee, a touch of lime juice, and a garnish of cilantro.

Insanely Crispy Gluten-Free Buffalo Chicken Wings

Most chicken wings are very unhealthy—feedlot animal parts fried in unhealthy oils and breaded with refined flour (yuck!). These, however, are baked, grain-free, and full of nutrients. Chicken skin is full of collagen, as are the cartilage-rich joints of the chicken wing. Collagen consists of important amino acids that have become relatively rare in the modern diet. Note: some hot sauces contain garbage ingredients. When picking the hot sauce to use, ensure that it has only red pepper, vinegar, salt, and garlic.

SERVES 2–3

What you'll need:

Softened or melted coconut oil
1 pound organic free-range chicken wings
Garlic salt (I like Redmond Real Salt Organic Garlic Salt)
1/2 cup hot sauce (I like Frank's RedHot Original Cayenne Pepper Sauce)
2 tablespoons grass-fed butter
Extra cayenne pepper, optional

What to do:

1. Preheat the oven to 250°F and grease a baking sheet with coconut oil.
2. Place the wings on the prepared sheet and sprinkle with garlic salt. Give them a nice even seasoning (one side is fine).
3. Bake the wings for 45 minutes. Why such a low temperature? It helps dry out the wings and melts away extra fat and connective tissue. Very important! (Note: the wings are *not done* after this step—*do not eat yet*!)
4. Turn the heat up to 425°F and bake for another 45 minutes. Wings should have a nice golden brown color when done and will have shrunk

considerably. Remove from the oven and let sit at room temperature for 5 minutes.

5. While the wings are resting, combine the hot sauce and butter (adding extra cayenne if desired) in a small saucepan over very low heat just to warm the hot sauce and melt the butter.
6. Whisk the wing sauce, then transfer to a large bowl or pot. Throw in the wings and toss well to coat them with the sauce. Eat!

How to serve it:

1. I highly recommend serving these wings alongside a large salad, roasted veggies, or other plant matter.

Turmeric-Almond Chicken FINGERS

Who doesn't love chicken tenders? In this recipe, dreamed up by chef Liana Werner-Gray (author of *The Earth Diet*), almond flour and turmeric make a great crust that not only helps you stay away from grains and traditional breading but also provides a delicious means of integrating turmeric. You can also make chicken nuggets instead of tenders—just cut the chicken into square nugget shapes. Kids love it! (I've named these fingers after the FINGER study, which you may recall from page 18.)

SERVES 2–3

What you'll need:

3/4 cup extra-virgin coconut oil
1 egg
1 pound organic, free-range, boneless, skinless chicken breast, cut into strips (or use chicken tenders to save time)
1 cup almond flour

1½ tablespoons ground turmeric
1 teaspoon salt
Dash of freshly ground black pepper

What to do:

1. In a large skillet, heat the oil over medium-high heat.
2. While the oil is heating, in a large bowl, beat the egg, add the chicken, and toss to coat.
3. In a small bowl, whisk together the almond flour, turmeric, salt, and pepper. Spread the mixture onto a plate.
4. Remove the chicken strips from the egg mixture and dip them into the almond flour mixture. Coat them well all over.
5. Test the oil by dropping in a pinch of almond flour; when it sizzles, it's ready. Drop the chicken strips into the pan and cook for 4 to 5 minutes on each side, until golden brown and the chicken is cooked through.
6. When finished, place on top of paper towels to drain excess oil.

How to serve it:

1. These would go great with Sautéed Greens or the "Cheesy" Kale Salad.

Sautéed Greens

I'm always sautéing up some greens. They make an excellent bed for any of the dishes I've presented here. Once you add the kale or whatever dark, leafy green you're using, keep the pan covered so that the water that evaporates will actually help to steam the kale.

SERVES 2–3

What you'll need:

- 2 tablespoons extra-virgin olive oil
- 1 onion, chopped
- 4 cloves garlic, crushed and peeled
- 1 bunch kale, center ribs and stems removed, leaves torn or chopped
- 1/4 teaspoon salt
- 1/4 teaspoon freshly ground black pepper

What to do:

1. In a large skillet, heat the oil over medium heat. Throw in the onion and cook until softened, 4 to 5 minutes. Add the garlic and cook for 1 to 2 minutes, until aromatic. Add the kale, salt, and pepper, reduce the heat to medium-low, cover, and cook, stirring a few times, until softened (approximately 10 minutes).

How to serve it:

1. I love to add a grass-fed beef patty, a piece of wild salmon, 2 or 3 poached or lightly fried eggs, or a few chicken legs to these greens.

Better Brain Bowl

This is a super-simple recipe (if you can even call it that) that provides incredible brain nourishment in the form of monounsaturated fat, lutein, zeaxanthin, omega-3s, and fiber.

SERVES 1

What you'll need:

1 4.4-ounce can of sardines (I love Wild Planet Wild Sardines in Extra-Virgin Olive Oil with Lemon)
1 avocado
1 lemon wedge
1 tablespoon Primal Kitchen Chipotle Lime Mayo, optional

What to do:

1. Empty the can of sardines into a bowl. Slice up the avocado, add it to the bowl, and squeeze the lemon on top. If you want to kick things up a notch, add the Chipotle Lime Mayo!

"Cheesy" Kale Salad

Here's a delicious salad that's easy to make and savory enough to convert even the most salad-phobic of the bunch.

SERVES 2–3

What you'll need:

1 bunch kale, center ribs and stems removed (reserve these for juicing or eating later)
2 tablespoons extra-virgin olive oil
2 tablespoons apple cider vinegar
1/2 green bell pepper, chopped
1/4 cup nutritional yeast
1 teaspoon garlic powder
3/4 teaspoon salt

What to do:

1. Tear the kale leaves into small pieces and place them in a large bowl. Add the oil and vinegar and stir or massage it into the leaves to start to soften them. Add the green pepper, then the nutritional yeast, garlic powder, and salt and toss until everything is well combined.

How to serve it:

1. Eat as is, or mix in some anchovies. Or throw a grass-fed beef patty on top!

Brain-Boosting Raw Chocolate

Dark chocolate has been in the research journals a lot of late for its cognition-boosting effects. To construct a sugar-free recipe, I enlisted my good friend Tero Isokauppila. Tero is the founder of the mushroom company Four Sigmatic, but he's also one of the most knowledgeable people I know on cacao, the main ingredient in chocolate.

SERVES 3–4

What you'll need:

1 cup finely chopped cacao butter
1 cup extra-virgin coconut oil
2 tablespoons sugar-free sweetener of choice (I recommend monk fruit, erythritol, or stevia)
1/2 teaspoon vanilla powder
Pinch of sea salt
3 packets Four Sigmatic Lion's Mane Elixir (or 1 heaping teaspoon of lion's mane extract), optional
1 cup unsweetened raw cacao powder, plus more if needed

What to do:

1. Put the cacao butter in a double boiler or heatproof bowl set over a pan of just-simmering water (make sure the bowl doesn't touch the water and keep it over low heat; this is important for preserving the enzymes and brain-nourishing properties of the cacao). Stir until completely melted. Add the coconut oil and use a whisk or milk frother to combine until the fats are emulsified. Add the sweetener, vanilla powder, salt, and lion's mane, if using. Whisk again to combine.
2. Slowly add the cacao powder to the mixture until it reaches the consistency of thick cream, adding more if needed.
3. Pour the mixture into ice cube trays and place in the freezer for 30 to 60 minutes to harden. Let them soften for 5 to 10 minutes after taking them out of the freezer before serving.

Supplements

Fish Oil (EPA/DHA)

Fish oil, how do I love thee? Let me count the ways. A high-quality fish oil supplement is an abundant and practical source of preformed EPA and DHA omega-3 fats, and adding fish oil to your diet could be one of the single most powerful steps you can take for your brain's health and function. I take my fish oil with me whenever I travel and only skip taking it on days that I consume fatty fish. One important consideration: always look at the amount of EPA and DHA, *not* the total amount of oil. For example, if your supplement contains 1000 milligrams of fish oil and a relatively small proportion of EPA and DHA, you have a low-quality supplement.

Recommendation: Look to get around 500 milligrams DHA and 1000 milligrams EPA triglycerides daily from fish oil or from fatty fish in your diet. Refrigerate your fish oil to keep it fresh.

WHAT DO WHALES KNOW THAT WE DON'T? FISH OIL VS. KRILL OIL

Fatty acids bound into triglycerides make up the vast configuration of fats found in the body. But cell membranes, including those of neurons, are made of phospholipids rather than triglycerides. While many fish oil supplements provide DHA and EPA omega-3s in the form of triglycerides, the omega-3s provided by krill oil are found in the membrane-equivalent phospholipid form (krill oil is made from the tiny invertebrate crustaceans that make up much of a whale's diet).

While the majority of research validating the use of omega-3 supplementation for brain health and function has used fish oil, new research suggests that krill oil may provide a superior and more bioavailable form of omega-3, and in particular DHA, which is more readily absorbed and incorporated into neuronal membranes. Krill oil also contains a number of other vital nutrients such as choline and astaxanthin. The former is the precursor to the neurotransmitter acetylcholine, which is critical for optimal memory function, and the latter is a powerful fat-soluble antioxidant.

So, should you take krill oil instead of fish oil? The most sensible solution is to consume wild fish, which contains both triglyceride and phospholipid forms of EPA and DHA. Fish roe (caviar, or for sushi fans, *ikura* or *tobiko*) is also a delicious source of phospholipid-bound omega-3s. If cost is not an issue and you choose to supplement, covering your bases with both a triglyceride-based fish oil as well as krill oil may be beneficial. If cost or practicality is an issue, high-quality triglyceride fish oil should work fine.

Vitamin D_3

A recent meta-analysis has found that of all environmental risk factors for developing dementia, the evidence pinpointing low vitamin D was the strongest. Vitamin D deficiency may also impair your brain's ability to synthesize serotonin from its precursor tryptophan, leading to lowered levels of this neurotransmitter in the brain. This can lead to depression and brain fog.

The main source of vitamin D comes from our exposure to the sun's UVB rays. Today, many of us spend a lot of our time indoors, and our skin's exposure to the sun is limited—meaning we're likely to have lower levels of vitamin D. There are also many person-to-person differences that affect our ability to synthesize vitamin D. Young people produce more vitamin D than older people—for example, a seventy-year-old makes four times less vitamin D from the sun than a twenty-year-old. Those with darker skin pigments also produce less vitamin D. (Melanin, which gives brown skin its color, is evolution's natural sunscreen.) That means that if you're a person of color living in a northern latitude, supplementation may be particularly important.

Those who are overweight have less available vitamin D, because as a fat-soluble vitamin, it gets stored in fat tissue. This occurs with other fat-soluble vitamins as well (like vitamin E) and may explain why overweight and obese people are more likely to be deficient in vitamin D even with the same sun exposure as their leaner counterparts. Perhaps it's not a coincidence that three-quarters of US teens and adults are estimated to be deficient in vitamin D, paralleling the widespread obesity epidemic.

VITAMIN D: THE ANTI-AGING VITAMIN?

We've evolved in the sun, and vitamin D is a chemical workhorse that our biology came to count on. It plays a role in regulating the expression of nearly one thousand genes in the human body—that's nearly 5 percent of the human genome! It could almost be considered a wonder vitamin, except that vitamin D isn't even a true vitamin—it's a hormone dependent upon sun exposure.

Some of vitamin D's many duties involve dampening the pro-inflammatory response and defending your cells from the wear and tear of aging. In fact, women who had blood levels within 40 to 60 ng/ml were shown to have the longest *telomeres* compared to age-matched controls. Telomeres are structures that protect your DNA from damage, and they typically shorten with age. It is believed that having longer telomeres at any given age is better.

Another study found that in female identical twins, those twins with the lowest levels of vitamin D had shorter telomeres, corresponding to five years of accelerated biological aging. This certainly helps to understand whether "healthy" aging is a matter of nature (your genetics) or nurture (your environment). These women had the *same* nature (the same genetic makeup), but those with lower vitamin D looked biologically older under the microscope!

If supplementing, just remember: it's possible to have too much vitamin D in your blood. Vitamin D increases the absorption of calcium, and the major risk of vitamin D toxicity is *hypercalcemia*, or too much calcium in the blood (see vitamin K_2 below). This can lead to problems like artery calcification and kidney stones. On the other hand, it's impossible to get too much vitamin D from the sun—just remember to take proper sun precautions and not to burn.

While there is not a consensus on the ideal level of vitamin D, keeping blood levels in the range of 40 to 60 ng/ml seems to confer the lowest rate of all-cause mortality over a given time period, which includes nonaccidental death by any cause. Your doctor can easily check your levels with a routine blood draw. Insufficiency, as currently described by the Endocrine Society (which actually considers vitamin D's broader importance to the body distinct from bone health), is below 30 ng/ml.

Recommendation: 2000 to 5000 IU of vitamin D_3 per day, checked every six months by a doctor to ensure levels between 40 and 60 ng/ml.

Folate, Vitamin B_{12}, Vitamin B_6

The complex of vitamins known as the B vitamins includes vitamin B_9 or folate, and vitamin B_{12} or cobalamin. B_{12} is important for normal nerve function and preventing anemia (a deficiency of red blood cells). Folate, as I mentioned when discussing the virtues of dark leafy greens, is also an important part of something called the *methylation cycle*. Ensuring adequate folate (and B_{12}) helps keep homocysteine, a toxic amino acid, low. Your homocysteine level is easily determined by your physician with a simple blood test, but elevated homocysteine is common, affecting up to 30 percent of persons over sixty-five worldwide.[1]

Having elevated homocysteine has been linked not only with worse cognitive performance, but with a twofold risk for developing dementia, heart attack, and stroke. Odds of brain shrinkage are up to ten times higher in patients with high homocysteine compared to those with normal levels.[2] Consuming a vitamin B complex including folate, B_{12}, and B_6 can keep levels within the normal, healthy range.

Many people already unknowingly supplement with folate—this is because it is added to a wide array of foods including bread and multivitamins in the form of folic acid. Unfortunately, because of a common genetic mutation known as *MTHFR* (short for *methylenetetrahydrofolate reductase*), many people do not convert folic acid, which is synthetic, to active folate, called *methylfolate*. This can drive levels of homocysteine up, among other potential problems.

When supplementing with B vitamins, avoid taking megadoses, which is not necessary and yet common in supplements. Taking too much folate if you're deficient in B_{12} can actually *accelerate* brain aging, whereas having optimal amounts of both can have the desired, protective effect. One way to ensure a healthy balance is simply to eat foods rich in natural sources of folate—vegetables—and match this consumption with consumption of egg yolks, beef, chicken, salmon, or sardines, all rich sources of B_{12}.

Recommendation: Try to get your B vitamins from food. Have your doctor check levels of folate and B_{12}, along with homocysteine. If B levels are low, or homocysteine is high (below 9 umol/l s ideal; lower generally is better), consider supplementing. Start at 400 micrograms folate (as methylfolate or methyltetrahydrofolate), 500 micrograms B_{12} (methylcobalamin), and 20 milligrams B_6 daily.

Vitamin K_2

Vitamin K_2 is an essential nutrient. It is involved in calcium homeostasis, ensuring that the mineral stays in places we want it (like our bones and teeth) and doesn't build up in places we don't (like in our arteries and kidneys). Many people, including some doctors, confuse K_2 with K_1, a vitamin involved in clotting. But while deficiency in K_1 is rare and easy to spot by the excessive bleeding and bruising it causes, K_2 deficits may be more common and, unfortunately, manifest invisibly. Intake of vitamin K_2 has also been linked with reduced cancer incidence, increased insulin sensitivity, better brain health, and more.

Recommendation: 50 to 100 micrograms K_2 MK-7 daily.

Turmeric

Turmeric is a root used in Ayurvedic cooking for millennia. It has two compounds of note: curcumin, a polyphenol, which has demonstrated anti-inflammatory capabilities, and aromatic turmerone, which may help boost stem cells in the brain. I encourage you to use turmeric in your cooking and supplement as needed for pain or inflammatory conditions.

Recommendation: 500 to 1000 milligrams turmeric as needed. Ensure the formulation contains piperine (black pepper extract), which enhances bioavailability. Turmeric rhizome extracts or turmeric phytosome are purported to be formulations with higher bioavailability.

Astaxanthin

Astaxanthin is a carotenoid commonly found in krill oil and is what gives wild salmon and pink flamingos their reddish appearance. Though the research is somewhat limited for this little-known antioxidant, there is enough to warrant its inclusion in my daily regimen. Astaxanthin has been shown to have whole-body benefits, including boosting cognitive function, protecting the skin from sun damage, enhancing skin appearance, protecting the eyes, reducing inflammation, converting blood lipids to a more cardioprotective profile, providing potent antioxidant effects and free radical scavenging, and more. Some of these benefits appear to be mediated by its ability to upregulate genes that protect against DNA damage and the stresses of aging, including *FOX03*. I take this daily. Like other carotenoids, astaxanthin is fat soluble, so make sure to take it with food containing fat.

Recommendation: 12 milligrams daily with a fat-containing meal or snack.

Probiotics

The research on probiotics is new and evolving. I enjoy probiotic foods (such as kimchi and kombucha), but taking a probiotic supplement as well can't hurt, particularly if foods containing probiotics aren't appealing to you.

Recommendation: If you choose to supplement, look for one with a high number of different strains (the gut contains hundreds of different species!), and 5 to 10 billion colony-forming units (CFUs). Ensuring that you take your probiotic with a source of prebiotic fiber can help the organisms better "take hold" in the gritty and competitive gut environment.

ACKNOWLEDGMENTS

Max

So many people have lent their time, intellect, talents, and skills to helping me pull this book together that I can't possibly thank them all. But I can certainly try.

First and foremost, thank you to all of the researchers around the globe who are doing the science, proving that our choices matter when it comes to our cognitive performance and long-term brain health. I especially want to thank the countless experts who've gotten on the phone with me, welcomed me into their labs, Skyped with me, and answered my questions over e-mail. In particular: Robert Krikorian, Miia Kivipelto, Agnes Flöel, Suzanne de la Monte, Alessio Fasano, Lisa Mosconi, Mary Newport, Melissa Schilling, Nina Teicholz, James DiNicolantonio, and Felice Jacka. Also thank you to the institutions who have welcomed me: NYU Langone Medical Center, Harvard University, Brown University, Weill Cornell Medicine/NewYork-Presbyterian and the Alzheimer's Prevention Clinic, the Karolinska Institutet, and Charité Hospital.

A major thank-you to Richard Isaacson, my mentor, colleague, and friend. I have learned much about science from you. I'm grateful whenever I get to collaborate in your research and look forward to future endeavors together. (That includes spin class.)

Thanks to my literary agent, Giles Anderson: your guidance through this process has been invaluable.

The Harper Wave team: you guys are so, so awesome. I'm so happy to have worked together on this book. Karen, you are luminous. Sarah, thank you for editing these words. I'm beyond proud of what we've accomplished together.

Paul Grewal, thank you for contributing your invaluable time and expertise to my book. I couldn't have picked a better or more brilliant collaborator.

Mehmet Oz, Ali Perry, and the entire Dr. Oz team. I think it's the coolest thing ever to be a "core expert" on the show, and I wear the badge with honor.

Craig and Sarah Clemens, so much love for you guys! Craig, thanks for naming my baby (this book) and sharing your talents to help it make an impact. I look forward to our next karaoke jam.

Kristin Loberg, thank you for your brilliant and generous feedback during my writing process. You are an inspiration. I still owe you a yoga session.

The producers of *The Doctors* TV show, thank you for repeatedly having me on and letting me feed your hosts weird health trends. I do it with the best of intentions!

To my friends in health and wellness, thank you for welcoming me into your communities with support and inspiration: David and Leize Perlmutter, Mark Hyman, William Davis, Terry Wahls, Mary Newport, Emily Fletcher, Kelly LeVeque, Mike Mutzel, Erin Matlock, James Maskell, Alex Doman, Mark Sisson, Pedram Shojai, Steven Gundry, Maria Shriver, and the Digital Natives team.

Writing a book takes a tremendous amount of work and support. Other friends who have lent valuable insight, feedback, suggestions, or comments, or who just supported me during moments of doubt (of which there were many!): Liana Werner-Gray, Tero Isokauppila, Michele Promaulayko, Crosby Tailor, Mary Shenouda, Amanda Cole, Kendall Dabaghi, Noah Berman, Misha Hyman, Mike Berman, Alex Kip, Chris Gartin, Ryan Star, Hilla Medalia, Rachel Beider, James Swanwick, Alexandra Calma, Sean Carey, Dhru Purohit, Andrew Luer, Nariman Hamed, and Matt Bilinsky.

If I've forgotten your name, I'm sorry—reach out to me and I'll have you over for dinner.

A big thanks to every single person who follows me on Facebook, Twitter, and Instagram for continually inspiring me every day to continue in the pursuit of truth. I am humbled by your messages. I also want to give a shout-out to members of The Cortex, my Facebook group and street team, as well as anyone who is on my e-mail list. And of course, a major thank-you to anyone who contributed to the crowdfunding campaign for my documentary *Bread Head* (www.breadheadmovie.com), which began it all. Thank you, thank you, thank you for your support. It means the world to me.

Finally, thanks to my two brothers, Andrew and Benny, my dad, Bruce, my mom, Kathy, and Delilah.

Paul

I'd like to first acknowledge my grandmother, Jaspal Kaur, whose long struggle with Alzheimer's disease belied the power of her incredible brain and spirit. From being essentially orphaned in Macau to single-handedly starting the first coed primary school in her region of India, she was a pioneer and trailblazer. If the strategies outlined in this book can prevent what happened to her from happening to just one person, our efforts will have been more than well spent.

Thank you, Mom, for imprinting us with a fraction of your genius. Dad, I'm glad your mother kicked you out of India for selling your textbooks to buy racing pigeons.

Alex, Rikki, Sean, Jim, Upkar, thanks for your feedback and friendship.

Max, it's been an honor to simultaneously get to know you and collaborate with you on this project; it's a once-in-a-lifetime opportunity and I won't soon forget the leap of faith you took to include me in such a deeply personal endeavor.

RESOURCES

JOIN THE CORTEX, A PRIVATE FACEBOOK COMMUNITY

http://maxl.ug/thecortex

Have a question about anything in this book? The first place you should go is The Cortex. This is a private Facebook community I've created for people who are going through their own health journeys to share tips, tricks, recipes, research, and more. Many of them are experienced and follow the Genius Plan, while others are just starting out. Make sure to introduce yourself!

WATCH MY DOCUMENTARY, *BREAD HEAD*

www.breadheadmovie.com

My story is documented in my film, *Bread Head*, the first and only feature-length documentary solely about dementia prevention, because changes begin in the brain *decades* before the first symptom of memory loss. Check out the website to watch the film, see a trailer, find local screenings, and become a Bread Head activist.

JOIN MY OFFICIAL NEWSLETTER

www.maxlugavere.com

Want research broken down and delivered straight to your inbox? My newsletter is where I regularly share research articles (with easy-to-read summaries), impromptu interviews, and other easily digestible tidbits designed to improve your life. No spam *ever*—don't worry, I got your back!

Research Resources

One of the top ways that you can ensure the information you're getting is sound is to make sure the places you're looking are credible, and as close to the science as possible. These are *the only* sources I can recommend using to track and search scientific research:

SCIENCEDAILY

www.sciencedaily.com

This site republishes university press releases that often accompany study publications. It brings together research from all different disciplines, but you can often find good stuff here by scrolling down to Health News or clicking Health in the menu bar at the top.

Note: press releases from universities are not necessarily perfect, but they are a great starting place and usually provide links to the research discussed. Reading both the press release *and* the study paper can help you learn how to interpret research. And the releases are often the very sources that journalists will use to write their articles. So in essence, this site takes you straight to the source!

MEDICAL XPRESS

www.medicalxpress.com

This site does the same as ScienceDaily but is exclusively medical/health-related.

EUREKALERT!

www.eurekalert.com

This is similar to the above two resources—it publishes press releases—but is run by the American Association for the Advancement of Science, which publishes the scientific journal *Science*.

PUBMED

www.ncbi.nlm.nih.gov/pubmed

When researching, I often use PubMed. One way to use Google to search PubMed is to add "site:nih.gov" into your Google search. For example, "alzheimer's insulin site:nih.gov" would search the NIH website (which includes PubMed) for all articles mentioning Alzheimer's and insulin.

Product Resources

Want to know the exact brand of blue-blocking glasses I use? Or my favorite online meditation course? Or an easy way to get the highest-quality meat air-shipped to you monthly no matter where in the world you are? I've got your back. Over the years I have become friendly with many food producers, supplement companies, and products. Anything I recommend is something I've vetted and personally use. To check out my recommendations for specific products alluded to in this book, visit http://maxl.ug/GFresources.

Contact

Contact the authors for speaking, coaching, or just say hi!

MAX LUGAVERE

www.maxlugavere.com
info@maxlugavere.com
Facebook: facebook.com/maxlugavere
Twitter: twitter.com/maxlugavere
Instagram: instagram.com/maxlugavere

DR. PAUL GREWAL

www.mymd.nyc
Twitter: twitter.com/paulgrewalmd
Instagram: instagram.com/paulgrewalmd

NOTES

CHAPTER 1: THE INVISIBLE PROBLEM

1. Claire T. McEvoy et al., "Neuroprotective Diets Are Associated with Better Cognitive Function: The Health and Retirement Study," *Journal of the American Geriatrics Society* 65, no. 8 (2017).
2. P. Eriksson et al., "Neurogenesis in the Adult Human Hippocampus," *Nature Medicine* 4, no. 11 (1998): 1313–17.
3. John Westfall, James Mold, and Lyle Fagnan, "Practice-Based Research—'Blue Highways' on the NIH Roadmap," *Journal of the American Medical Association* 297, no. 4 (2007): 403–6.
4. O. Rogowski et al., "Waist Circumference as the Predominant Contributor to the Micro-Inflammatory Response in the Metabolic Syndrome: A Cross Sectional Study," *Journal of Inflammation* 26 (2010): 35.
5. NCD Risk Factor Collaboration, "Trends in Adult Body-Mass Index in 200 Countries from 1975 to 2014: A Pooled Analysis of 1698 Population-based Measurement Studies with 19.2 Million Participants," *Lancet* 387, no. 10026 (2016): 1377–96.
6. Jeffrey Blumberg et al., "Vitamin and Mineral Intake Is Inadequate for Most Americans: What Should We Advise Patients About Supplements?" supplement to *Journal of Family Practice* 65, no. 9 (2016): S1–8.

GENIUS FOOD #1: EXTRA-VIRGIN OLIVE OIL

1. Michael Hopkin, "Extra-Virgin Olive Oil Mimics Painkiller," *Nature*, August 31, 2005, http://www.nature.com/drugdisc/news/articles/050829-11.html.
2. A. Abuznait et al., "Olive-Oil-Derived Oleocanthal Enhances B-Amyloid Clearance as a Potential Neuroprotective Mechanism against Alzheimer's Disease: In Vitro and In Vivo Studies," *ACS Chemical Neuroscience* 4, no. 6 (2013): 973–82.
3. E. H. Martinez-Lapiscina et al., "Mediterranean Diet Improves Cognition: The PREDIMED-NAVARRA Randomised Trial," *Journal of Neurology, Neurosurgery, and Psychiatry* 84, no. 12 (2013): 1318–25.
4. J. A. Menendez et al., "Analyzing Effects of Extra-Virgin Olive Oil Polyphenols on Breast Cancer-Associated Fatty Acid Synthase Protein Expression Using Reverse-Phase Protein Microarrays," *International Journal of Molecular Medicine* 22, no. 4 (2008): 433–39.

CHAPTER 2: FANTASTIC FATS AND OMINOUS OILS

1. Antonio Gotto Jr., "Evolving Concepts of Dyslipidemia, Atherosclerosis, and Cardiovascular Disease: The Louis F. Bishop Lecture," *Journal of the American College of Cardiology* 46, no. 7 (2005): 1219–24.

2. Ian Leslie, "The Sugar Conspiracy," *Guardian*, April 7, 2016, http://www.theguardian.com/society/2016/apr/07/the-sugar-conspiracy-robert-lustig-john-yudkin?CMP=share_btn_tw.
3. Cristin Kearns, Laura Schmidt, and Stanton Glantz, "Sugar Industry and Coronary Heart Disease Research: A Historical Analysis of Internal Industry Documents," *JAMA Internal Medicine* 176, no. 11 (2016): 1680–85.
4. Anahad O'Connor, "Coca-Cola Funds Scientists Who Shift Blame for Obesity Away from Bad Diets," *New York Times*, August 9, 2015, https://well.blogs.nytimes.com/2015/08/09/coca-cola-funds-scientists-who-shift-blame-for-obesity-away-from-bad-diets/?_r=0.
5. L. Lluis et al., "Protective Effect of the Omega-3 Polyunsaturated Fatty Acids: Eicosapentaenoic Acid/Docosahexaenoic Acid 1:1 Ratio on Cardiovascular Disease Risk Markers in Rats," *Lipids in Health and Disease* 12, no. 140 (2013): 140.
6. National Cancer Institute, "Table 2. Food Sources of Total Omega 6 Fatty Acids (18:2 + 20:4), Listed in Descending Order by Percentages of Their Contribution to Intake, Based on Data from the National Health and Nutrition Examination Survey 2005–2006," https://epi.grants.cancer.gov/diet/foodsources/fatty_acids/table2.html.
7. K. Chen, M. Kazachkov, and P. H. Yu, "Effect of Aldehydes Derived from Oxidative Deamination and Oxidative Stress on B-Amyloid Aggregation; Pathological Implications to Alzheimer's Disease," *Journal of Neural Transmission* 114 (2007): 835–39.
8. R. A. Vaishnav et al., "Lipid Peroxidation-Derived Reactive Aldehydes Directly and Differentially Impair Spinal Cord and Brain Mitochondrial Function," *Journal of Neurotrauma* 27, no. 7 (2010): 1311–20.
9. G. Spiteller and M. Afzal, "The Action of Peroxyl Radicals, Powerful Deleterious Reagents, Explains Why Neither Cholesterol nor Saturated Fatty Acids Cause Atherogenesis and Age-Related Diseases," *Chemistry* 20, no. 46 (2014): 14298–345.
10. T. L. Blasbalg et al., "Changes in Consumption of Omega-3 and Omega-6 Fatty Acids in the United States During the 20th Century," *American Journal of Clinical Nutrition* 93, no. 5 (2011): 950–62.
11. Sean O'Keefe et al., "Levels of Trans Geometrical Isomers of Essential Fatty Acids in Some Unhydrogenated US Vegetable Oils," *Journal of Food Lipids* 1, no. 3 (1994): 165–76.
12. A. P. Simopoulos, "Evolutionary Aspects of Diet: The Omega-6/Omega-3 Ratio and the Brain," *Molecular Neurobiology* 44, no. 2 (2011): 203–15.
13. Janice Kiecolt-Glaser et al., "Omega-3 Supplementation Lowers Inflammation and Anxiety in Medical Students: A Randomized Controlled Trial," *Brain, Behavior, and Immunity* 25, no. 8 (2011): 1725–34.
14. Lon White et al., "Prevalence of Dementia in Older Japanese-American Men in Hawaii: The Honolulu-Asia Aging Study," *Journal of the American Medical Association* 276, no. 12 (1996): 955–60.
15. D. S. Heron et al., "Lipid Fluidity Markedly Modulates the Binding of Serotonin to Mouse Brain Membranes," *Proceedings of the National Academy of Sciences* 77, no. 12 (1980): 7463–67.
16. A. Veronica Witte et al., "Long-Chain Omega-3 Fatty Acids Improve Brain Function and Structure in Older Adults," *Cerebral Cortex* 24, no. 11 (2014): 3059–68; Aaron T. Piepmeier and Jennifer L. Etnier, "Brain-Derived Neurotrophic Factor (BDNF) as a Po-

tential Mechanism of the Effects of Acute Exercise on Cognitive Performance," *Journal of Sport and Health Science* 4, no. 1 (2015): 14–23.

17. Paul S. Aisen, "Serum Brain-Derived Neurotrophic Factor and the Risk for Dementia," *Journal of the American Medical Association* 311, no. 16 (2014): 1684–85.
18. Bun-Hee Lee and Yong-Ku Kim, "The Roles of BDNF in the Pathophysiology of Major Depression and in Antidepressant Treatment," *Psychiatry Investigation* 7, no. 4 (2010): 231–35.
19. James V. Pottala et al., "Higher RBC EPA + DHA Corresponds with Larger Total Brain and Hippocampal Volumes: WHIMS-MRI Study," *Neurology* 82, no. 5 (2014): 435–42.
20. Ellen Galinsky, "Executive Function Skills Predict Children's Success in Life and in School," *Huffington Post*, June 21, 2012, http://www.huffingtonpost.com/ellen-galinsky/executive-function-skills_1_b_1613422.html.
21. Kelly Sheppard and Carol Cheatham, "Omega-6 to Omega-3 Fatty Acid Ratio and Higher-Order Cognitive Functions in 7- to 9-year-olds: A Cross-Sectional Study," *American Journal of Clinical Nutrition* 98, no. 3 (2013): 659–67.
22. M. H. Bloch and A. Qawasmi, "Omega-3 Fatty Acid Supplementation for the Treatment of Children with Attention-Deficit/Hyperactivity Disorder Symptomatology: Systematic Review and Meta-Analysis," *Journal of the American Academy of Child Adolescent Psychiatry* 50, no. 10 (2011): 991–1000; D. J. Bos et al., "Reduced Symptoms of Inattention after Dietary Omega-3 Fatty Acid Supplementation in Boys with and without Attention Deficit/Hyperactivity Disorder," *Neuropsychopharmacology* 40, no. 10 (2015): 2298–306.
23. Witte, "Long-Chain Omega-3 Fatty Acids."
24. G. Paul Amminger et al., "Longer-Term Outcome in the Prevention of Psychotic Disorders by the Vienna Omega-3 Study," *Nature Communications* 6 (2015).
25. Christine Wendlinger and Walter Vetter, "High Concentrations of Furan Fatty Acids in Organic Butter Samples from the German Market," *Journal of Agricultural and Food Chemistry* 62, no. 34 (2014): 8740–44.
26. D. F. Horrobin, "Loss of Delta-6-Desaturase Activity as a Key Factor in Aging," *Medical Hypotheses* 7, no. 9 (1981): 1211–20.
27. Tamas Decsi and Kathy Kennedy, "Sex-Specific Differences in Essential Fatty Acid Metabolism," *American Journal of Clinical Nutrition* 94, no. 6 (2011): 1914S–19S.
28. R. A. Mathias et al., "Adaptive Evolution of the FADS Gene Cluster within Africa," *PLOS ONE* 7, no. 9 (2012): e44926.
29. Y. Allouche et al., "How Heating Affects Extra-Virgin Olive Oil Quality Indexes and Chemical Composition," *Journal of Agricultural and Food Chemistry* 55, no. 23 (2007): 9646–54; S. Casal et al., "Olive Oil Stability under Deep-Frying Conditions," *Food and Chemical Toxicology* 48, no. 10 (2010): 2972–79.
30. Sara Staubo et al., "Mediterranean Diet, Micronutrients and Macronutrients, and MRI Measures of Cortical Thickness," *Alzheimer's & Dementia* 13, no. 2 (2017): 168–77.
31. Cinta Valls-Pedret et al., "Mediterranean Diet and Age-Related Cognitive Decline," *JAMA Internal Medicine* 175, no. 7 (2015): 1094–103.
32. W. M. Fernando et al., "The Role of Dietary Coconut for the Prevention and Treatment of Alzheimer's Disease: Potential Mechanisms of Action," *British Journal of Nutrition* 114, no. 1 (2015): 1–14; B. Jarmolowska et al., "Changes of Beta-Casomorphin Content in Human Milk During Lactation," *Peptides* 28, no. 10 (2007): 1982–86.

33. Euridice Martinez Steele et al., "Ultra-Processed Foods and Added Sugars in the US Diet: Evidence from a Nationally Representative Cross-Sectional Study," *BMJ Open* 6 (2016).
34. Camille Amadieu et al., "Nutrient Biomarker Patterns and Long-Term Risk of Dementia in Older Adults," *Alzheimer's & Dementia* 13, no. 10 (2017).
35. Brittanie M. Volk et al., "Effects of Step-wise Increases in Dietary Carbohydrate on Circulating Saturated Fatty Acids and Palmitoleic Acid in Adults with Metabolic Syndrome," *PLOS ONE* 9, no. 11 (2014): e113605.
36. Cassandra Forsythe et al., "Comparison of Low Fat and Low Carbohydrate Diets on Circulating Fatty Acid Composition and Markers of Inflammation," *Lipids* 43, no. 1 (2008): 65–77.
37. Felice Jacka et al., "Western Diet Is Associated with a Smaller Hippocampus: A Longitudinal Investigation," *BMC Medicine* 13 (2015): 215.
38. A. Wu et al., "A Saturated-Fat Diet Aggravates the Outcome of Traumatic Brain Injury on Hippocampal Plasticity and Cognitive Function by Reducing Brain-Derived Neurotrophic Factor," *Neuroscience* 119, no. 2 (2003): 365–75.
39. David DiSalvo, "How a High-Fat Diet Could Damage Your Brain," Forbes.com, November 30, 2015, http://www.forbes.com/sites/daviddisalvo/2015/11/30/how-a-high-fat-diet-could-damage-your-brain/#2f784e59661c.
40. G. L. Bowman et al., "Nutrient Biomarker Patterns, Cognitive Function, and MRI Measures of Brain Aging," *Neurology* 78, no. 4 (2011).
41. Beatrice Golomb, "A Fat to Forget: Trans Fat Consumption and Memory," *PLOS ONE* 10, no. 6 (2015).
42. Marta Zamroziewicz et al., "Parahippocampal Cortex Mediates the Relationship between Lutein and Crystallized Intelligence in Healthy, Older Adults," *Frontiers in Aging Neuroscience* 8 (2016).
43. M. J. Brown et al., "Carotenoid Bioavailability Is Higher from Salads Ingested with Full-Fat than with Fat-Reduced Salad Dressings as Measured with Electromechanical Detection," *American Journal of Clinical Nutrition* 80, no. 2 (2004): 396–403.
44. Amy Patterson Neubert, "Study: Top Salads with Eggs to Better Absorb Vegetables' Carotenoids," Purdue University, June 4, 2015, http://www.purdue.edu/newsroom/releases/2015/Q2/study-top-salads-with-eggs-to-better-absorb-vegetables-carotenoids-.html.

CHAPTER 3: OVERFED, YET STARVING

1. Loren Cordain et al., "Plant-Animal Subsistence Ratios and Macronutrient Energy Estimations in Worldwide Hunter-Gatherer Diets," *American Journal of Clinical Nutrition* 71, no. 3 (2000): 682–92.
2. Steele, "Ultra-Processed Foods."
3. Blumberg, "Vitamin and Mineral Intake."
4. Lewis Killin et al., "Environmental Risk Factors for Dementia: A Systematic Review," *BMC Geriatrics* 16 (2016): 175.
5. Creighton University, "Recommendation for Vitamin D Intake Was Miscalculated, Is Far Too Low, Experts Say," ScienceDaily, March 17, 2015, https://www.sciencedaily.com/releases/2015/03/150317122458.htm.
6. A. Rosanoff, C. M. Weaver, and R. K. Rude, "Suboptimal Magnesium Status in the United States: Are the Health Consequences Underestimated?" *Nutrition Review* 70, no. 3 (2012): 153–64.

7. Pauline Anderson, "Inflammatory Dietary Pattern Linked to Brain Aging," Medscape, July 17, 2017, https://www.medscape.com/viewarticle/883038.
8. Timothy Lyons, "Glycation and Oxidation of Proteins: A Role in the Pathogenesis of Atherosclerosis," in *Drugs Affecting Lipid Metabolism* (Kluwer Academic Publishers, 1993), 407–20.
9. J. Uribarri et al., "Circulating Glycotoxins and Dietary Advanced Glycation Endproducts: Two Links to Inflammatory Response, Oxidative Stress, and Aging," *Journals of Gerontology, Series A: Biological Sciences and Medical Sciences* 62, no. 4 (2007): 427–33.
10. P. I. Moreira et al., "Oxidative Stress and Neurodegeneration," *Annals of the New York Academy of Sciences* 1043 (2005): 545–52.
11. N. Sasaki et al., "Advanced Glycation End Products in Alzheimer's Disease and Other Neurodegenerative Diseases," *American Journal of Pathology* 153, no. 4 (1998): 1149–55.
12. M. S. Beeri et al., "Serum Concentration of an Inflammatory Glycotoxin, Methylglyoxal, Is Associated with Increased Cognitive Decline in Elderly Individuals," *Mechanisms of Ageing and Development* 132, no. 11–12 (2011): 583–87; K. Yaffe et al., "Advanced Glycation End Product Level, Diabetes, and Accelerated Cognitive Aging," *Neurology* 77, no. 14 (2011): 1351–56; Weijing Cai et al., "Oral Glycotoxins Are a Modifiable Cause of Dementia and the Metabolic Syndrome in Mice and Humans," *Proceedings of the National Academy of Sciences* 111, no. 13 (2014): 4940–45.
13. American Academy of Neurology, "Lower Blood Sugars May Be Good for the Brain," ScienceDaily, October 23, 2013, https://www.sciencedaily.com/releases/2013/10/131023165016.htm.
14. American Academy of Neurology, "Even in Normal Range, High Blood Sugar Linked to Brain Shrinkage," ScienceDaily, September 4, 2012, https://www.sciencedaily.com/releases/2012/09/120904095856.htm.
15. Mark A. Virtue et al., "Relationship between GHb Concentration and Erythrocyte Survival Determined from Breath Carbon Monoxide Concentration," *Diabetes Care* 27, no. 4 (2004): 931–35.
16. C. Luevano-Contreras and K. Chapman-Novakofski, "Dietary Advanced Glycation End Products and Aging," *Nutrients* 2, no. 12 (2010): 1247–65.
17. S. Swamy-Mruthinti et al., "Evidence of a Glycemic Threshold for the Development of Cataracts in Diabetic Rats," *Current Eye Research* 18, no. 6 (1999): 423–29.
18. N. G. Rowe et al., "Diabetes, Fasting Blood Glucose and Age-Related Cataract: The Blue Mountains Eye Study," *Ophthalmic Epidemiology* 7, no. 2 (2000): 106–14.
19. M. Krajcovicova-Kudlackova et al., "Advanced Glycation End Products and Nutrition," *Physiological Research* 51, no. 2 (2002): 313–16.
20. Nicole J. Kellow et al., "Effect of Dietary Prebiotic Supplementation on Advanced Glycation, Insulin Resistance and Inflammatory Biomarkers in Adults with Pre-diabetes: A Study Protocol for a Double-Blind Placebo-Controlled Randomised Crossover Clinical Trial," *BMC Endocrine Disorders* 14, no. 1 (2014): 55.
21. V. Lecoultre et al., "Effects of Fructose and Glucose Overfeeding on Hepatic Insulin Sensitivity and Intrahepatic Lipids in Healthy Humans," *Obesity (Silver Spring)* 21, no. 4 (2013): 782–85.
22. Qingying Meng et al., "Systems Nutrigenomics Reveals Brain Gene Networks Linking Metabolic and Brain Disorders," *EBioMedicine* 7 (2016): 157–66.
23. Do-Geun Kim et al., "Non-alcoholic Fatty Liver Disease Induces Signs of Alzheimer's

Disease (AD) in Wild-Type Mice and Accelerates Pathological Signs of AD in an AD Model," *Journal of Neuroinflammation* 13 (2016).

24. M. Ledochowski et al., "Fructose Malabsorption Is Associated with Decreased Plasma Tryptophan," *Scandinavian Journal of Gastroenterology* 36, no. 4 (2001): 367–71.
25. M. Ledochowski et al., "Fructose Malabsorption Is Associated with Early Signs of Mental Depression," *European Journal of Medical Research* 17, no. 3 (1998): 295–98.
26. Shannon L. Macauley et al., "Hyperglycemia Modulates Extracellular Amyloid-β Concentrations and Neuronal Activity in Vivo," *Journal of Clinical Investigation* 125, no. 6 (2015): 2463.
27. Paul K. Crane et al., "Glucose Levels and Risk of Dementia," *New England Journal of Medicine* 2013, no. 369 (2013): 540–48.
28. Derrick Johnston Alperet et al., "Influence of Temperate, Subtropical, and Tropical Fruit Consumption on Risk of Type 2 Diabetes in an Asian Population," *American Journal of Clinical Nutrition* 105, no. 3 (2017).
29. Y. Gu et al., "Mediterranean Diet and Brain Structure in a Multiethnic Elderly Cohort," *Neurology* 85, no. 20 (2015): 1744–51.
30. Staubo, "Mediterranean Diet."
31. E. E. Devore et al., "Dietary Intakes of Berries and Flavonoids in Relation to Cognitive Decline," *Annals of Neurology* 72, no. 1 (2012): 135–43.
32. Martha Clare Morris et al., "MIND Diet Associated with Reduced Incidence of Alzheimer's Disease," *Alzheimer's & Dementia* 11, no. 9 (2015): 1007–14.
33. O'Connor, "Coca-Cola Funds Scientists."
34. Christopher J. L. Murray et al., "The State of US Health, 1990–2010: Burden of Diseases, Injuries, and Risk Factors," *Journal of the American Medical Association* 310, no. 6 (2013): 591–606.
35. Susan Jones, "11,774 Terror Attacks Worldwide in 2015; 28,328 Deaths Due to Terror Attacks," CNSNews.com, June 3, 2016, http://www.cnsnews.com/news/article/susan-jones/11774-number-terror-attacks-worldwide-dropped-13-2015.
36. Robert Proctor, "The History of the Discovery of the Cigarette–Lung Cancer Link: Evidentiary Traditions, Corporate Denial, Global Toll," *Tobacco Control* 21, no. 2 (2011): 87–91.

GENIUS FOOD #3: BLUEBERRIES

1. C. M. Williams et al., "Blueberry-Induced Changes in Spatial Working Memory Correlate with Changes in Hippocampal CREB Phosphorylation and Brain-Derived Neurotrophic Factor (BDNF) Levels," *Free Radical Biological Medicine* 45, no. 3 (2008): 295–305.
2. R. Krikorian et al., "Blueberry Supplementation Improves Memory in Older Adults," *Journal of Agricultural Food Chemistry* 58, no. 7 (2010): 3996–4000.
3. Elizabeth Devore et al., "Dietary Intakes of Berries and Flavonoids in Relation to Cognitive Decline," *Annals of Neurology* 72, no. 1 (2012): 135–43.
4. M. C. Morris et al., "MIND Diet Slows Cognitive Decline with Aging," *Alzheimer's & Dementia* 11, no. 9 (2015): 1015–22.

CHAPTER 4: WINTER IS COMING (FOR YOUR BRAIN)

1. K. de Punder and L. Pruimboom, "The Dietary Intake of Wheat and Other Cereal Grains and Their Role in Inflammation," *Nutrients* 5, no. 3 (2013): 771–87.

2. Ibid.
3. J. R. Kraft and W. H. Wehrmacher, "Diabetes—A Silent Disorder," *Comprehensive Therapy* 35, nos. 3–4 (2009): 155–59.
4. Jean-Sebastien Joyal et al., "Retinal Lipid and Glucose Metabolism Dictates Angiogenesis through the Lipid Sensor Ffar1," *Nature Medicine* 22, no. 4 (2016): 439–45.
5. Chung-Jung Chiu et al., "Dietary Carbohydrate and the Progression of Age-Related Macular Degeneration: A Prospective Study from the Age-Related Eye Disease Study," *American Journal of Clinical Nutrition* 86, no. 4 (2007): 1210–18.
6. Matthew Harber et al., "Alterations in Carbohydrate Metabolism in Response to Short-Term Dietary Carbohydrate Restriction," *American Journal of Physiology—Endocrinology and Metabolism* 289, no. 2 (2005): E306–12.
7. Brian Morris et al., "FOX03: A Major Gene for Human Longevity—A Mini-Review," *Gerontology* 61, no. 6 (2015): 515–25.
8. Ibid.
9. Valerie Renault et al., "FOX03 Regulates Neural Stem Cell Homeostasis," *Cell Stem Cell* 5 (2009): 527–39.
10. J. M. Bao et al., "Association between FOX03A Gene Polymorphisms and Human Longevity: A Meta-Analysis," *Asian Journal of Andrology* 16, no. 3 (2014): 446–52.
11. Brian Morris, "FOX03: A Major Gene for Human Longevity."
12. Catherine Crofts et al., "Hyperinsulinemia: A Unifying Theory of Chronic Disease?" *Diabesity* 1, no. 4 (2015): 34–43.
13. W. Q. Qui et al., "Insulin-Degrading Enzyme Regulates Extracellular Levels of Amyloid Beta-Protein by Degradation," *Journal of Biological Chemistry* 273, no. 49 (1998): 32730–38.
14. Y. M. Li and D. W. Dickson, "Enhanced Binding of Advanced Glycation Endproducts (AGE) by the ApoE4 Isoform Links the Mechanism of Plaque Deposition in Alzheimer's Disease," *Neuroscience Letters* 226, no. 3 (1997): 155–58.
15. Auriel Willette et al., "Insulin Resistance Predicts Brain Amyloid Deposition in Late Middle-Aged Adults," *Alzheimer's & Dementia* 11, no. 5 (2015): 504–10.
16. L. P. van der Heide et al., "Insulin Modulates Hippocampal Activity-Dependent Synaptic Plasticity in a N-Methyl-D-Aspartate Receptor and Phosphatidyl-Inositol-3-Kinase-Dependent Manner," *Journal of Neurochemistry* 94, no. 4 (2005): 1158–66.
17. H. Bruehl et al., "Cognitive Impairment in Nondiabetic Middle-Aged and Older Adults Is Associated with Insulin Resistance," *Journal of Clinical and Experimental Neuropsychology* 32, no. 5 (2010): 487–93.
18. Kaarin Anstey et al., "Association of Cognitive Function with Glucose Tolerance and Trajectories of Glucose Tolerance over 12 Years in the AusDiab Study," *Alzheimer's Research & Therapy* 7, no. 1 (2015): 48; S. E. Young, A. G. Mainous 3rd, and M. Carnemolla, "Hyperinsulinemia and Cognitive Decline in a Middle-Aged Cohort," *Diabetes Care* 29, no. 12 (2006): 2688–93.
19. B. Kim and E. L. Feldman, "Insulin Resistance as a Key Link for the Increased Risk of Cognitive Impairment in the Metabolic Syndrome," *Exploratory Molecular Medicine* 47 (2015): e149.
20. Dimitrios Kapogiannis et al., "Dysfunctionally Phosphorylated Type 1 Insulin Receptor Substrate in Neural-Derived Blood Exosomes of Preclinical Alzheimer's Disease," *FASEB Journal* 29, no. 2 (2015): 589–96.
21. G. Collier and K. O'Dea, "The Effect of Coingestion of Fat on the Glucose, Insulin,

and Gastric Inhibitory Polypeptide Responses to Carbohydrate and Protein," *American Journal of Clinical Nutrition* 37, no. 6 (1983): 941–44.

22. Sylvie Normand et al., "Influence of Dietary Fat on Postprandial Glucose Metabolism (Exogenous and Endogenous) Using Intrinsically C-Enriched Durum Wheat," *British Journal of Nutrition* 86, no. 1 (2001): 3–11.
23. M. Sorensen et al., "Long-Term Exposure to Road Traffic Noise and Incident Diabetes: A Cohort Study," *Environmental Health Perspectives* 121, no. 2 (2013): 217–22.
24. R. H. Freire et al., "Wheat Gluten Intake Increases Weight Gain and Adiposity Associated with Reduced Thermogenesis and Energy Expenditure in an Animal Model of Obesity," *International Journal of Obesity* 40, no. 3 (2016): 479–87; Fabíola Lacerda Pires Soares et al., "Gluten-Free Diet Reduces Adiposity, Inflammation and Insulin Resistance Associated with the Induction of PPAR-Alpha and PPAR-Gamma Expression," *Journal of Nutritional Biochemistry* 24, no. 6 (2013): 1105–11.
25. Thi Loan Anh Nguyen et al., "How Informative Is the Mouse for Human Gut Microbiota Research?" *Disease Models & Mechanisms* 8, no. 1 (2015): 1–16.
26. Matthew S. Tryon et al., "Excessive Sugar Consumption May Be a Difficult Habit to Break: A View from the Brain and Body," *Journal of Clinical Endocrinology & Metabolism* 100, no. 6 (2015): 2239–47.
27. Marcia de Oliveira Otto et al., "Everything in Moderation—Dietary Diversity and Quality, Central Obesity and Risk of Diabetes," *PLOS ONE* 10, no. 10 (2015).
28. Sarah A. M. Kelly et al., "Whole Grain Cereals for the Primary or Secondary Prevention of Cardiovascular Disease," *The Cochrane Library* (2017).

GENIUS FOOD #4: DARK CHOCOLATE

1. Adam Brickman et al., "Enhancing Dentate Gyrus Function with Dietary Flavanols Improves Cognition in Older Adults," *Nature Neuroscience* 17, no. 12 (2014): 1798–803.
2. Georgina Crichton, Merrill Elias, and Ala'a Alkerwi, "Chocolate Intake Is Associated with Better Cognitive Function: The Maine-Syracuse Longitudinal Study," *Appetite* 100 (2016): 126–32.

CHAPTER 5: HEALTHY HEART, HEALTHY BRAIN

1. M. L. Alosco et al., "The Adverse Effects of Reduced Cerebral Perfusion on Cognition and Brain Structure in Older Adults with Cardiovascular Disease," *Brain Behavior* 3, no. 6 (2013): 626–36.
2. P. W. Siri-Tarino et al., "Meta-Analysis of Prospective Cohort Studies Evaluating the Association of Saturated Fat with Cardiovascular Disease," *American Journal of Clinical Nutrition* 91, no. 3 (2010): 535–46.
3. I. D. Frantz Jr. et al., "Test of Effect of Lipid Lowering by Diet on Cardiovascular Risk. The Minnesota Coronary Survey," *Arteriosclerosis* 9, no. 1 (1989): 129–35.
4. Christopher Ramsden et al., "Re-evaluation of the Traditional Diet-Heart Hypothesis: Analysis of Recovered Data from Minnesota Coronary Experiment (1968–73)," *BMJ* 353 (2016); Anahad O'Connor, "A Decades-Old Study, Rediscovered, Challenges Advice on Saturated Fat," *New York Times*, April 13, 2016, https://well.blogs.nytimes.com/2016/04/13/a-decades-old-study-rediscovered-challenges-advice-on-saturated-fat/.
5. Matthias Orth and Stefano Bellosta, "Cholesterol: Its Regulation and Role in Central Nervous System Disorders," *Cholesterol* (2012).

6. P. K. Elias et al., "Serum Cholesterol and Cognitive Performance in the Framingham Heart Study," *Psychosomatic Medicine* 67, no. 1 (2005): 24–30.
7. R. West et al., "Better Memory Functioning Associated with Higher Total and Low-Density Lipoprotein Cholesterol Levels in Very Elderly Subjects without the Apolipoprotein e4 Allele," *American Journal of Geriatric Psychiatry* 16, no. 9 (2008): 781–85.
8. B. G. Schreurs, "The Effects of Cholesterol on Learning and Memory," *Neuroscience & Biobehavioral Reviews* 34, no. 8 (2010): 1366–79; M. M. Mielke et al., "High Total Cholesterol Levels in Late Life Associated with a Reduced Risk of Dementia," *Neurology* 64, no. 10 (2005): 1689–95.
9. Credit Suisse, "Credit Suisse Publishers Report on Evolving Consumer Perceptions about Fat," PR Newswire, September 17, 2015, http://www.prnewswire.com/news-releases/credit-suisse-publishes-report-on-evolving-consumer-perceptions-about-fat-300144839.html.
10. Marja-Leena Silaste et al., "Changes in Dietary Fat Intake Alter Plasma Levels of Oxidized Low-Density Lipoprotein and Lipoprotein(a)," *Arteriosclerosis, Thrombosis, and Vascular Biology* 24, no. 3 (2004): 495–503.
11. Patty W. Siri-Tarino et al., "Saturated Fatty Acids and Risk of Coronary Heart Disease: Modulation by Replacement Nutrients," *Current Atherosclerosis Reports* 12, no. 6 (2010): 384–90.
12. V. A. Mustad et al., "Reducing Saturated Fat Intake Is Associated with Increased Levels of LDL Receptors on Mononuclear Cells in Healthy Men and Women," *Journal of Lipid Research* 38, no. 3 (March 1997): 459–68.
13. L. Li et al., "Oxidative LDL Modification Is Increased in Vascular Dementia and Is Inversely Associated with Cognitive Performance," *Free Radical Research* 44, no. 3 (2010): 241–48.
14. Steen G. Hasselbalch et al., "Changes in Cerebral Blood Flow and Carbohydrate Metabolism during Acute Hyperketonemia," *American Journal of Physiology—Endocrinology and Metabolism* 270, no. 5 (1996): E746–51.
15. E. L. Wightman et al., "Dietary Nitrate Modulates Cerebral Blood Flow Parameters and Cognitive Performance in Humans: A Double-Blind, Placebo-Controlled, Crossover Investigation," *Physiological Behavior* 149 (2015): 149–58.
16. Riaz Memon et al., "Infection and Inflammation Induce LDL Oxidation In Vivo," *Arteriosclerosis, Thrombosis, and Vascular Biology* 20 (2000): 1536–42.
17. A. C. Vreugdenhil et al., "LPS-Binding Protein Circulates in Association with ApoB-Containing Lipoproteins and Enhances Endotoxin-LDL/VLDL Interaction," *Journal of Clinical Investigation* 107, no. 2 (2001): 225–34.
18. B. M. Charalambous et al., "Role of Bacterial Endotoxin in Chronic Heart Failure: The Gut of the Matter," *Shock* 28, no. 1 (2007): 15–23.
19. Stephen Bischoff et al., "Intestinal Permeability—A New Target for Disease Prevention and Therapy," *BMC Gastroenterology* 14 (2014): 189.
20. C. U. Choi et al., "Statins Do Not Decrease Small, Dense Low-Density Lipoprotein," *Texas Heart Institute Journal* 37, no. 4 (2010): 421–28.
21. Melinda Wenner Moyer, "It's Not Dementia, It's Your Heart Medication: Cholesterol Drugs and Memory," *Scientific American*, September 1, 2010, https://www.scientificamerican.com/article/its-not-dementia-its-your-heart-medication/.
22. "Coenzyme Q10," Linus Pauling Institute—Macronutrient Information Center, Oregon State University, http://lpi.oregonstate.edu/mic/dietary-factors/coenzyme-Q10.

23. I. Mansi et al., "Statins and New-Onset Diabetes Mellitus and Diabetic Complications: A Retrospective Cohort Study of US Healthy Adults," *Journal of General Internal Medicine* 30, no. 11 (2015): 1599–610.
24. Shannon Macauley et al., "Hyperglycemia Modulates Extracellular Amyloid-B Concentrations and Neuronal Activity In Vivo," *Journal of Clinical Investigation* 125, no. 6 (2015): 2463–67.

GENIUS FOOD #5: EGGS

1. C. N. Blesso et al., "Whole Egg Consumption Improves Lipoprotein Profiles and Insulin Sensitivity to a Greater Extent than Yolk-Free Egg Substitute in Individuals with Metabolic Syndrome," *Metabolism* 62, no. 3 (2013): 400–10.
2. Garry Handelman et al., "Lutein and Zeaxanthin Concentrations in Plasma after Dietary Supplementation with Egg Yolk," *American Journal of Clinical Nutrition* 70, no. 2 (1999): 247–51.

CHAPTER 6: FUELING YOUR BRAIN

1. L. Kovac, "The 20 W Sleep-Walkers," *EMBO Reports* 11, no. 1 (2010): 2.
2. NCD Risk Factor Collaboration, "Trends in Adult Body-Mass Index."
3. Institute for Basic Science, "Team Suppresses Oxidative Stress, Neuronal Death Associated with Alzheimer's Disease," ScienceDaily, February 25, 2016, https://www.sciencedaily.com/releases/2016/02/160225085645.htm.
4. J. Ezaki et al., "Liver Autophagy Contributes to the Maintenance of Blood Glucose and Amino Acid Levels," *Autophagy* 7, no. 7 (2011): 727–36.
5. H. White and B. Venkatesh, "Clinical Review: Ketones and Brain Injury," *Critical Care* 15, no. 2 (2011): 219.
6. R. L. Veech et al., "Ketone Bodies, Potential Therapeutic Uses," *IUBMB Life* 51, no. 4 (2001): 241–47.
7. S. G. Jarrett et al., "The Ketogenic Diet Increases Mitochondrial Glutathione Levels," *Journal of Neurochemistry* 106, no. 3 (2008): 1044–51.
8. Sama Sleiman et al., "Exercise Promotes the Expression of Brain Derived Neurotrophic Factor (BDNF) through the Action of the Ketone Body β-Hydroxybutyrate," *Cell Biology* (2016).
9. Hasselbalch, "Changes in Cerebral Blood Flow."
10. Jean-Jacques Hublin and Michael P. Richards, eds., *The Evolution of Hominin Diets: Integrating Approaches to the Study of Palaeolithic Subsistence* (Springer Science & Business Media, 2009).
11. S. T. Henderson, "Ketone Bodies as a Therapeutic for Alzheimer's Disease," *Neurotherapeutics* 5, no. 3 (2008): 470–80.
12. S. Brandhorst et al., "A Periodic Diet that Mimics Fasting Promotes Multi-System Regeneration, Enhanced Cognitive Performance, and Healthspan," *Cell Metabolism* 22, no. 1 (2016): 86–99.
13. Caroline Rae et al., "Oral Creatine Monohydrate Supplementation Improves Brain Performance: A Double-Blind, Placebo-Controlled, Cross-over Trial," *Proceedings of the Royal Society of London B: Biological Sciences* 270, no. 1529 (2003): 2147–50.
14. J. Delanghe et al., "Normal Reference Values for Creatine, Creatinine, and Carnitine Are Lower in Vegetarians," *Clinical Chemistry* 35, no. 8 (1989): 1802–3.
15. Rafael Deminice et al., "Creatine Supplementation Reduces Increased Homocysteine

Concentration Induced by Acute Exercise in Rats," *European Journal of Applied Physiology* 111, no. 11 (2011): 2663–70.

16. David Benton and Rachel Donohoe, "The Influence of Creatine Supplementation on the Cognitive Functioning of Vegetarians and Omnivores," *British Journal of Nutrition* 105, no. 7 (2011): 1100–1105.
17. Rachel N. Smith, Amruta S. Agharkar, and Eric B. Gonzales, "A Review of Creatine Supplementation in Age-Related Diseases: More than a Supplement for Athletes," *F1000Research* 3 (2014).
18. Terry McMorris et al., "Creatine Supplementation and Cognitive Performance in Elderly Individuals," *Aging, Neuropsychology, and Cognition* 14, no. 5 (2007): 517–28.
19. M. P. Laakso et al., "Decreased Brain Creatine Levels in Elderly Apolipoprotein E ε4 Carriers," *Journal of Neural Transmission* 110, no. 3 (2003): 267–75.
20. A. L. Rogovik and R. D. Goldman, "Ketogenic Diet for Treatment of Epilepsy," *Canadian Family Physician* 56, no. 6 (2010): 540–42.
21. Zhong Zhao et al., "A Ketogenic Diet as a Potential Novel Therapeutic Intervention in Amyotrophic Lateral Sclerosis," *BMC Neuroscience* 7, no. 29 (2006).
22. R. Krikorian et al., "Dietary Ketosis Enhances Memory in Mild Cognitive Impairment," *Neurobiology of Aging* 425, no. 2 (2012): 425e19–27; Matthew Taylor et al., "Feasibility and efficacy data from a ketogenic diet intervention in Alzheimer's disease," *Alzheimer's & Dementia: Translational Research and Clinical Interventions* (2017).
23. S. Djiogue et al., "Insulin Resistance and Cancer: The Role of Insulin and IGFs," *Endocrine-Related Cancer* 20, no. 1 (2013): R1–17.
24. Harber, "Alterations in Carbohydrate Metabolism."
25. Heikki Pentikäinen et al., "Muscle Strength and Cognition in Ageing Men and Women: The DR's EXTRA Study," *European Geriatric Medicine* 8 (2017).
26. Henderson, "Ketone Bodies as a Therapeutic."
27. E. M. Reiman et al., "Functional Brain Abnormalities in Young Adults at Genetic Risk for Late-Onset Alzheimer's Dementia," *Proceedings of the National Academy of Sciences USA* 101, no. 1 (2004): 284–89.
28. S. T. Henderson, "High Carbohydrate Diets and Alzheimer's Disease," *Medical Hypotheses* 62, no. 5 (2004): 689–700.
29. Hugh C. Hendrie et al., "APOE ε4 and the Risk for Alzheimer Disease and Cognitive Decline in African Americans and Yoruba," *International Psychogeriatrics* 26, no. 6 (2014): 977–985.
30. Henderson, "High Carbohydrate Diets."
31. Konrad Talbot et al., "Demonstrated Brain Insulin Resistance in Alzheimer's Disease Patients Is Associated with IGF-1 Resistance, IRS-1 Dysregulation, and Cognitive Decline," *Journal of Clinical Investigation* 122, no. 4 (2012).
32. Dale E. Bredesen, "Reversal of Cognitive Decline: A Novel Therapeutic Program," *Aging* 6, no. 9 (2014): 707.
33. S. C. Cunnane et al., "Can Ketones Help Rescue Brain Fuel Supply in Later Life? Implications for Cognitive Health during Aging and the Treatment of Alzheimer's Disease," *Frontiers in Molecular Neuroscience* 9 (2016): 53.
34. M. Gasior, M. A. Rogawski, and A. L. Hartman, "Neuroprotective and Disease-Modifying Effects of the Ketogenic Diet," *Behavioral Pharmacology* 17, nos. 5–6 (2006): 431–39.
35. S. L. Kesl et al., "Effects of Exogenous Ketone Supplementation on Blood Ketone, Glu-

cose, Triglyceride, and Lipoprotein Levels in Sprague-Dawley Rats," *Nutrition & Metabolism London* 13 (2016): 9.

36. W. Zhao et al., "Caprylic Triglyceride as a Novel Therapeutic Approach to Effectively Improve the Performance and Attenuate the Symptoms Due to the Motor Neuron Loss in ALS Disease," *PLOS ONE* 7, no. 11 (2012): e49191.
37. D. Mungas et al., "Dietary Preference for Sweet Foods in Patients with Dementia," *Journal of the American Geriatric Society* 38, no. 9 (1990): 999–1007.
38. M. A. Reger et al., "Effects of Beta-Hydroxybutyrate on Cognition in Memory-Impaired Adults," *Neurobiology of Aging* 25, no. 3 (2004): 311–14.

GENIUS FOOD #6: GRASS-FED BEEF

1. Janet R. Hunt, "Bioavailability of Iron, Zinc, and Other Trace Minerals from Vegetarian Diets," *American Journal of Clinical Nutrition* 78, no. 3 (2003): 633S–39S.
2. Felice N. Jacka et al., "Red Meat Consumption and Mood and Anxiety Disorders," *Psychotherapy and Psychosomatics* 81, no. 3 (2012): 196–98.
3. Charlotte G. Neumann et al., "Meat Supplementation Improves Growth, Cognitive, and Behavioral Outcomes in Kenyan Children," *Journal of Nutrition* 137, no. 4 (2007): 1119–23.
4. Shannon P. McPherron et al., "Evidence for Stone-Tool-Assisted Consumption of Animal Tissues before 3.39 Million Years Ago at Dikika, Ethiopia," *Nature* 466, no. 7308 (2010): 857–60.
5. M. Gibis, "Effect of Oil Marinades with Garlic, Onion, and Lemon Juice on the Formation of Heterocyclic Aromatic Amines in Fried Beef Patties," *Journal of Agricultural Food Chemistry* 55, no. 25 (2007): 10240–47.
6. Wataru Yamadera et al., "Glycine Ingestion Improves Subjective Sleep Quality in Human Volunteers, Correlating with Polysomnographic Changes," *Sleep and Biological Rhythms* 5, no. 2 (2007): 126–31; Makoto Bannai et al., "Oral Administration of Glycine Increases Extracellular Serotonin but Not Dopamine in the Prefrontal Cortex of Rats," *Psychiatry and Clinical Neurosciences* 65, no. 2 (2011): 142–49.

CHAPTER 7: GO WITH YOUR GUT

1. Camilla Urbaniak et al., "Microbiota of Human Breast Tissue," *Applied and Environmental Microbiology* 80, no. 10 (2014): 3007–14.
2. American Society for Microbiology, "Cities Have Individual Microbial Signatures," ScienceDaily, April 19, 2016, https://www.sciencedaily.com/releases/2016/04/160419144724.htm.
3. Ron Sender, Shai Fuchs, and Ron Milo, "Revised Estimates for the Number of Human and Bacteria Cells in the Body," *PLOS Biology* 14, no. 8 (2016): e1002533.
4. Mark Bowden, "The Measured Man," *Atlantic*, February 19, 2014, https://www.theatlantic.com/magazine/archive/2012/07/the-measured-man/309018/.
5. Robert A. Koeth et al., "Intestinal Microbiota Metabolism of L-Carnitine, a Nutrient in Red Meat, Promotes Atherosclerosis," *Nature Medicine* 19, no. 5 (2013): 576–85.
6. Jeff Leach, "From Meat to Microbes to Main Street: Is It Time to Trade In Your George Foreman Grill?" Human Food Project, April 18, 2013, http://www.humanfoodproject.com/from-meat-to-microbes-to-main-street-is-it-time-to-trade-in-your-george-foreman-grill/.

7. Francesca De Filippis et al., "High-Level Adherence to a Mediterranean Diet Beneficially Impacts the Gut Microbiota and Associated Metabolome," *Gut* 65, no. 11 (2015).
8. Roberto Berni Canani, Margherita Di Costanzo, and Ludovica Leone, "The Epigenetic Effects of Butyrate: Potential Therapeutic Implications for Clinical Practice," *Clinical Epigenetics* 4, no. 1 (2012): 4.
9. K. Meijer, P. de Vos, and M. G. Priebe, "Butyrate and Other Short-Chain Fatty Acids as Modulators of Immunity: What Relevance for Health?" *Current Opinion in Clinical Nutrition & Metabolic Care* 13, no. 6 (2010): 715–21.
10. A. L. Marsland et al., "Interleukin-6 Covaries Inversely with Cognitive Performance among Middle-Aged Community Volunteers," *Psychosomatic Medicine* 68, no. 6 (2006): 895–903.
11. Yasumichi Arai et al., "Inflammation, but Not Telomere Length, Predicts Successful Ageing at Extreme Old Age: A Longitudinal Study of Semi-supercentenarians," *EBioMedicine* 2, no. 10 (2015): 1549–58.
12. Christopher J. L. Murray et al., "Global, Regional, and National Disability-Adjusted Life Years (DALYs) for 306 Diseases and Injuries and Healthy Life Expectancy (HALE) for 188 Countries, 1990–2013: Quantifying the Epidemiological Transition," *Lancet* 386, no. 10009 (2015): 2145–91.
13. Bamini Gopinath et al., "Association between Carbohydrate Nutrition and Successful Aging over 10 Years," *Journals of Gerontology* 71, no. 10 (2016): 1335–40.
14. H. Okada et al., "The 'Hygiene Hypothesis' for Autoimmune and Allergic Diseases: An Update," *Clinical & Experimental Immunology* 160, no. 1 (2010): 1–9.
15. S. Y. Kim et al., "Differential Expression of Multiple Transglutaminases in Human Brain. Increased Expression and Cross-Linking by Transglutaminases 1 and 2 in Alzheimer's Disease," *Journal of Biological Chemistry* 274, no. 43 (1999): 30715–21; G. Andringa et al., "Tissue Transglutaminase Catalyzes the Formation of Alpha-Synuclein Crosslinks in Parkinson's Disease," *FASEB Journal* 18, no. 7 (2004): 932–34; A. Gadoth et al., "Transglutaminase 6 Antibodies in the Serum of Patients with Amyotrophic Lateral Sclerosis," *JAMA Neurology* 72, no. 6 (2015): 676–81.
16. C. L. Ch'ng et al., "Prospective Screening for Coeliac Disease in Patients with Graves' Hyperthyroidism Using Anti-gliadin and Tissue Transglutaminase Antibodies," *Clinical Endocrinology Oxford* 62, no. 3 (2005): 303–6.
17. Clare Wotton and Michael Goldacre, "Associations between Specific Autoimmune Diseases and Subsequent Dementia: Retrospective Record-Linkage Cohort Study, UK," *Journal of Epidemiology & Community Health* 71, no. 6 (2017).
18. C. L. Ch'ng, M. K. Jones, and J. G. Kingham, "Celiac Disease and Autoimmune Thyroid Disease," *Clinical Medicine Research* 5, no. 3 (2007): 184–92.
19. Julia Bollrath and Fiona Powrie, "Feed Your T_{regs} More Fiber," *Science* 341, no. 6145 (2013): 463–64.
20. Paola Bressan and Peter Kramer, "Bread and Other Edible Agents of Mental Disease," *Frontiers in Human Neuroscience* 10 (2016).
21. Alessio Fasano, "Zonulin, Regulation of Tight Junctions, and Autoimmune Diseases," *Annals of the New York Academy of Sciences* 1258, no. 1 (2012): 25–33.
22. R. Dantzer et al., "From Inflammation to Sickness and Depression: When the Immune System Subjugates the Brain," *Nature Reviews Neuroscience* 9, no. 1 (2008): 46–56.
23. A. H. Miller, V. Maletic, and C. L. Raison, "Inflammation and Its Discontents: The

Role of Cytokines in the Pathophysiology of Major Depression," *Biological Psychiatry* 65, no. 9 (2009): 732–41.

24. "Depression," World Health Organization, February 2017, http://www.who.int/mediacentre/factsheets/fs369/en/.
25. Alessio Fasano, "Zonulin and Its Regulation of Intestinal Barrier Function: The Biological Door to Inflammation, Autoimmunity, and Cancer," *Physiological Reviews* 91, no. 1 (2011): 151–75; E. Lionetti et al., "Gluten Psychosis: Confirmation of a New Clinical Entity," *Nutrients* 7, no. 7 (2015): 5532–39.
26. Melanie Uhde et al., "Intestinal Cell Damage and Systemic Immune Activation in Individuals Reporting Sensitivity to Wheat in the Absence of Coeliac Disease," *Gut* 65, no. 12 (2016).
27. Blaise Corthésy, H. Rex Gaskins, and Annick Mercenier, "Cross-talk between Probiotic Bacteria and the Host Immune System," *Journal of Nutrition* 137, no. 3 (2007): 781S–90S.
28. S. Bala et al., "Acute Binge Drinking Increases Serum Endotoxin and Bacterial DNA Levels in Healthy Individuals," *PLOS ONE* 9, no. 5 (2014): e96864.
29. V. Purohit et al., "Alcohol, Intestinal Bacterial Growth, Intestinal Permeability to Endotoxin, and Medical Consequences: A Summary of a Symposium," *Alcohol* 42, no. 5 (2008): 349–61.
30. Manfred Lamprecht and Anita Frauwallner, "Exercise, Intestinal Barrier Dysfunction and Probiotic Supplementation," *Acute Topics in Sport Nutrition* 59 (2012): 47–56.
31. Angela E. Murphy, Kandy T. Velazquez, and Kyle M. Herbert, "Influence of High-Fat Diet on Gut Microbiota: A Driving Force for Chronic Disease Risk," *Current Opinion in Clinical Nutrition and Metabolic Care* 18, no. 5 (2015): 515.
32. J. R. Rapin and N. Wiernsperger, "Possible Links between Intestinal Permeability and Food Processing: A Potential Therapeutic Niche for Glutamine," *Clinics Sao Paulo* 65, no. 6 (2010): 635–43.
33. E. Gaudier et al., "Butyrate Specifically Modulates MUC Gene Expression in Intestinal Epithelial Goblet Cells Deprived of Glucose," *American Journal of Physiology–Gastrointestinal and Liver Physiology* 287, no. 6 (2004): G1168–74.
34. Thi Loan Anh Nguyen et al., "How Informative Is the Mouse for Human Gut Microbiota Research?" *Disease Models & Mechanisms* 8, no. 1 (2015): 1–16.
35. Benoit Chassaing et al., "Dietary Emulsifiers Impact the Mouse Gut Microbiota Promoting Colitis and Metabolic Syndrome," *Nature* 519, no. 7541 (2015): 92–96.
36. Ian Sample, "Probiotic Bacteria May Aid Against Anxiety and Memory Problems," *Guardian*, October 18, 2015, https://www.theguardian.com/science/2015/oct/18/probiotic-bacteria-bifidobacterium-longum-1714-anxiety-memory-study.
37. Merete Ellekilde et al., "Transfer of Gut Microbiota from Lean and Obese Mice to Antibiotic-Treated Mice," *Scientific Reports* 4 (2014): 5922; Peter J. Turnbaugh et al., "An Obesity-Associated Gut Microbiome with Increased Capacity for Energy Harvest," *Nature* 444, no. 7122 (2006): 1027–131.
38. Kirsten Tillisch et al., "Brain Structure and Response to Emotional Stimuli as Related to Gut Microbial Profiles in Healthy Women," *Psychosomatic Medicine* 79, no. 8 (2017).
39. Giada De Palma et al., "Transplantation of Fecal Microbiota from Patients with Irritable Bowel Syndrome Alters Gut Function and Behavior in Recipient Mice," *Science Translational Medicine* 9, no. 379 (2017): eaaf6397.
40. Leach, "From Meat to Microbes to Main Street"; Gary D. Wu et al., "Linking Long-Term Dietary Patterns with Gut Microbial Enterotypes," *Science* 334, no. 6052 (2011): 105–8.

41. Bruce Goldman, "Low-Fiber Diet May Cause Irreversible Depletion of Gut Bacteria over Generations," Stanford Medicine News Center, January 13, 2016, http://med.stanford.edu/news/all-news/2016/01/low-fiber-diet-may-cause-irreversible-depletion-of-gut-bacteria.html.
42. T. K. Schaffer et al., "Evaluation of Antioxidant Activity of Grapevine Leaves Extracts (*Vitis labrusca*) in Liver of Wistar Rats," *Anais da Academia Brasileria de Ciencias* 88, no. 1 (2016): 187–96; T. Taira et al., "Dietary Polyphenols Increase Fecal Mucin and Immunoglobulin A and Ameliorate the Disturbance in Gut Microbiota Caused by a High Fat Diet," *Journal of Clinical Biochemical Nutrition* 57, no. 3 (2015): 212–16.
43. Pranita Tamma and Sara Cosgrove, "Addressing the Appropriateness of Outpatient Antibiotic Prescribing in the United States," *Journal of the American Medical Association* 315, no. 17 (2016): 1839–41.
44. R. Dunn et al., "Home Life: Factors Structuring the Bacterial Diversity Found within and between Homes," *PLOS ONE* 8, no. 5 (2013): e64133; Uppsala Universitet, "Early Contact with Dogs Linked to Lower Risk of Asthma," ScienceDaily, November 2, 2015, https://www.sciencedaily.com/releases/2015/11/151102143636.htm.
45. M. Samsam, R. Ahangari, and S. A. Naser, "Pathophysiology of Autism Spectrum Disorders: Revisiting Gastrointestinal Involvement and Immune Imbalance," *World Journal of Gastroenterology* 20, no. 29 (2014): 9942–51.
46. Elisabeth Svensson et al., "Vagotomy and Subsequent Risk of Parkinson's Disease," *Annals of Neurology* 78, no. 4 (2015): 522–29.
47. Floyd Dewhirst et al., "The Human Oral Microbiome," *Journal of Bacteriology* 192, no. 19 (2010): 5002–17.
48. M. Ide et al., "Periodontitis and Cognitive Decline in Alzheimer's Disease," *PLOS ONE* 11, no. 3 (2016): e0151081.

CHAPTER 8: YOUR BRAIN'S CHEMICAL SWITCHBOARD

1. Uwe Rudolph, "GABAergic System," *Encyclopedia of Molecular Pharmacology*, 515–19.
2. William McEntee and Thomas Crook, "Glutamate: Its Role in Learning, Memory, and the Aging Brain," *Psychopharmacology* 111, no. 4 (1993): 391–401.
3. "Disease Mechanisms," ALS Association, accessed November 7, 2017, http://www.alsa.org/research/focus-areas/disease-mechanisms.
4. Javier A. Bravo et al., "Ingestion of *Lactobacillus* Strain Regulates Emotional Behavior and Central GABA Receptor Expression in a Mouse via the Vagus Nerve," *Proceedings of the National Academy of Sciences* 108, no. 38 (2011): 16050–55.
5. Expertanswer, "*Lactobacillus reuteri* Good for Health, Swedish Study Finds," Science Daily, November 4, 2010, https://www.sciencedaily.com/releases/2010/11/101102131302.htm.
6. Richard Maddock et al., "Acute Modulation of Cortical Glutamate and GABA Content by Physical Activity," *Journal of Neuroscience* 36, no. 8 (2016): 2449–57.
7. Eric Herbst and Graham Holloway, "Exercise Increases Mitochondrial Glutamate Oxidation in the Mouse Cerebral Cortex," *Applied Physiology, Nutrition, and Metabolism* 41, no. 7 (2016): 799–801.
8. Boston University, "Yoga May Elevate Brain GABA Levels, Suggesting Possible Treatment for Depression," ScienceDaily, May 22, 2007, https://www.sciencedaily.com/releases/2007/05/070521145516.htm.
9. T. M. Makinen et al., "Autonomic Nervous Function during Whole-Body Cold Ex-

posure before and after Cold Acclimation," *Aviation, Space, and Environmental Medicine* 79, no. 9 (2008): 875–82.

10. K. Rycerz and J. E. Jaworska-Adamu, "Effects of Aspartame Metabolites on Astrocytes and Neurons," *Folia Neuropathological* 51, no. 1 (2013): 10–17.
11. Xueya Cai et al., "Long-Term Anticholinergic Use and the Aging Brain," *Alzheimer's & Dementia* 9, no. 4 (2013): 377–85.
12. Shelly Gray et al., "Cumulative Use of Strong Anticholinergics and Incident Dementia: A Prospective Cohort Study," *JAMA Internal Medicine* 175, no. 3 (2015): 401–7.
13. Richard Wurtman, "Effects of Nutrients on Neurotransmitter Release," in *Food Components to Enhance Performance: An Evaluation of Potential Performance-Enhancing Food Components for Operational Rations,* ed. Bernadette M. Marriott (Washington, DC: National Academies Press, 1994).
14. Institute of Medicine, "Choline," in *Dietary Reference Intakes for Thiamin, Riboflavin, Niacin, Vitamin B_6, Folate, Vitamin B_{12}, Pantothenic Acid, Biotin, and Choline* (Washington, DC: National Academies Press, 1998).
15. Helen Jensen et al., "Choline in the Diets of the US Population: NHANES, 2003–2004," *FASEB Journal* 21 (2007): LB46.
16. Roland Griffiths et al., "Psilocybin Produces Substantial and Sustained Decreases in Depression and Anxiety in Patients with Life-Threatening Cancer," *Journal of Psychopharmacology* 30, no. 12 (2016).
17. S. N. Young, "Acute Tryptophan Depletion in Humans: A Review of Theoretical, Practical, and Ethical Aspects," *Journal of Psychiatry & Neuroscience* 38, no. 5 (2013): 294–305.
18. S. N. Young and M. Leyton, "The Role of Serotonin in Human Mood and Social Interaction. Insight from Altered Tryptophan Levels," *Pharmacology Biochemistry and Behavior* 71, no. 4 (2002): 857–65.
19. S. N. Young et al., "Bright Light Exposure during Acute Tryptophan Depletion Prevents a Lowering of Mood in Mildly Seasonal Women," *European Neuropsychopharmacology* 18, no. 1 (2008): 14–23.
20. R. P. Patrick and B. N. Ames, "Vitamin D and the Omega-3 Fatty Acids Control Serotonin Synthesis and Action, Part 2: Relevance for ADHD, Bipolar Disorder, Schizophrenia, and Impulsive Behavior," *FASEB Journal* 29, no. 6 (2015): 2207–22.
21. Roni Caryn Rabin, "A Glut of Antidepressants," *New York Times*, August 12, 2013, https://well.blogs.nytimes.com/2013/08/12/a-glut-of-antidepressants/?mcubz=0.
22. Jay Fournier et al., "Antidepressant Drug Effects and Depression Severity: A Patient-Level Meta-analysis," *Journal of the American Medical Association* 303, no. 1 (2010): 47–53.
23. Ibid.; A. L. Lopresti and P. D. Drummond, "Efficacy of Curcumin, and a Saffron /Curcumin Combination for the Treatment of Major Depression: A Randomised, Double-Blind, Placebo-Controlled Study," *Journal of Affective Disorders* 201 (2017): 188–96.
24. F. Chaouloff et al., "Motor Activity Increases Tryptophan, 5-Hydroxyindoleacetic Acid, and Homovanillic Acid in Ventricular Cerebrospinal Fluid of the Conscious Rat," *Journal of Neurochemistry* 46, no. 4 (1986): 1313–16.
25. Stephane Thobois et al., "Role of Dopaminergic Treatment in Dopamine Receptor

Down-Regulation in Advanced Parkinson Disease: A Positron Emission Tomographic Study," *JAMA Neurology* 61, no. 11 (2004): 1705–9.

26. Richard A. Friedman, "A Natural Fix for A.D.H.D.," *New York Times*, October 31, 2014, https://www.nytimes.com/2014/11/02/opinion/sunday/a-natural-fix-for-adhd.html?mcubz=0.
27. Matt McFarland, "Crazy Good: How Mental Illnesses Help Entrepreneurs Thrive," *Washington Post*, April 29, 2015, https://www.washingtonpost.com/news/innovations/wp/2015/04/29/crazy-good-how-mental-illnesses-help-entrepreneurs-thrive/?utm_term=.37b4bc5bc699.
28. P. Rada, N. M. Avena, and B. G. Hoebel, "Daily Bingeing on Sugar Repeatedly Releases Dopamine in the Accumbens Shell," *Neuroscience* 134, no. 3 (2005): 737–44.
29. Fengqin Liu et al., "It Takes Biking to Learn: Physical Activity Improves Learning a Second Language," *PLOS ONE* 12, no. 5 (2017): e0177624.
30. B. J. Cardinal et al., "If Exercise Is Medicine, Where Is Exercise in Medicine? Review of U.S. Medical Education Curricula for Physical Activity-Related Content," *Journal of Physical Activity and Health* 12, no. 9 (2015): 1336–45.
31. K. Kukkonen-Harjula et al., "Haemodynamic and Hormonal Responses to Heat Exposure in a Finnish Sauna Bath," *European Journal of Applied Physiology and Occupational Physiology* 58, no. 5 (1989): 543–50.
32. T. Laatikainen et al., "Response of Plasma Endorphins, Prolactin and Catecholamines in Women to Intense Heat in a Sauna," *European Journal of Applied Physiology and Occupational Physiology* 57, no. 1 (1988): 98–102.
33. P. Sramek et al., "Human Physiological Responses to Immersion into Water of Different Temperatures," *European Journal of Applied Physiology* 81, no. 5 (2000): 436–42.
34. McGill University, "Vulnerability to Depression Linked to Noradrenaline," EurekAlert!, February 15, 2016, https://www.eurekalert.org/pub_releases/2016-02/mu-vtd021216.php.
35. M. T. Heneka et al., "Locus Ceruleus Controls Alzheimer's Disease Pathology by Modulating Microglial Functions through Norepinephrine," *Proceedings of the National Academy of Sciences USA* 107, no. 13 (2010): 6058–63.
36. Ibid.
37. University of Southern California, "Researchers Highlight Brain Region as 'Ground Zero' of Alzheimer's Disease: Essential for Maintaining Cognitive Function as a Person Ages, the Tiny Locus Coeruleus Region of the Brain Is Vulnerable to Toxins and Infection," ScienceDaily, February 16, 2016, https://www.sciencedaily.com/releases/2016/02/160216142835.htm.
38. A. Samara, "Single Neurons Needed for Brain Asymmetry Studies," *Frontiers in Genetics* 16, no. 4 (2014): 311.
39. M. S. Parihar and G. J. Brewer, "Amyloid-β as a Modulator of Synaptic Plasticity," *Journal of Alzheimer's Disease* 22, no. 3 (2010): 741–63.
40. Ganesh Shankar and Dominic Walsh, "Alzheimer's Disease: Synaptic Dysfunction and Aβ," *Molecular Neurodegeneration* 4, no. 48 (2009).
41. Gianni Pezzoli and Emanuele Cereda, "Exposure to Pesticides or Solvents and Risk of Parkinson Disease," *Neurology* 80, no. 22 (2013): 2035–41.
42. T. P. Brown et al., "Pesticides and Parkinson's Disease—Is There a Link?" *Environmental Health Perspectives* 114, no. 2 (2006): 156–64.

43. Grant Kauwe et al., "Acute Fasting Regulates Retrograde Synaptic Enhancement through a 4E-BP-Dependent Mechanism," *Neuron* 92, no. 6 (2016): 1204–12.
44. Jonah Lehrer, "The Neuroscience of Inception," *Wired*, July 26, 2010, https://www wired.com/2010/07/the-neuroscience-of-inception/.
45. Steven James et al., "Hominid Use of Fire in the Lower and Middle Pleistocene: A Review of the Evidence," *Current Anthropology* 30, no. 1 (1989).

GENIUS FOOD #8: BROCCOLI

1. S. K. Ghawi, L. Methven, and K. Nlranjan, "The Potential to Intensity Sulforaphane Formation in Cooked Broccoli (*Brassica oleracea var. italica*) Using Mustard Seeds (*Sinapis alba*)," *Food Chemistry* 138, nos. 2–3 (2013): 1734–41.

CHAPTER 9: SACRED SLEEP (AND THE HORMONAL HELPERS)

1. J. Zhang et al., "Extended Wakefulness: Compromised Metabolics in and Degeneration of Locus Ceruleus Neurons," *Journal of Neuroscience* 34, no. 12 (2014): 4418–31.
2. C. Benedict et al., "Acute Sleep Deprivation Increases Serum Levels of Neuron-Specific Enolase (NSE) and S100 Calcium Binding Protein B (S-100B) in Healthy Young Men," *Sleep* 37, no. 1 (2014): 195–98.
3. National Sleep Foundation, "Bedroom Poll," accessed November 7, 2017, https://sleep foundation.org/sites/default/files/bedroompoll/NSF_Bedroom_Poll_Report.pdf.
4. American Psychological Association, "Stress in America: Our Health at Risk," January 11, 2012, http://www.apa.org/news/press/releases/stress/2011/final-2011.pdf.
5. A. P. Spira et al., "Self-Reported Sleep and β-amyloid Deposition in Community-Dwelling Older Adults," *JAMA Neurology* 70, no. 12 (2013): 1537–43.
6. Huixia Ren et al., "Omega-3 Polyunsaturated Fatty Acids Promote Amyloid-β Clearance from the Brain through Mediating the Function of the Glymphatic System," *FASEB Journal* 31, no. 1 (2016).
7. A. Afaghi, H. O'Connor, and C. M. Chow, "Acute Effects of the Very Low Carbohydrate Diet on Sleep Indices," *Nutritional Neuroscience* 11, no. 4 (2008): 146–54.
8. Marie-Pierre St-Onge et al., "Fiber and Saturated Fat Are Associated with Sleep Arousals and Slow Wave Sleep," *Journal of Clinical Sleep Medicine* 12, no. 1 (2016): 19–24.
9. Seung-Gul Kang et al., "Decrease in fMRI Brain Activation during Working Memory Performed after Sleeping under 10 Lux Light," *Scientific Reports* 6 (2016): 36731.
10. Cibele Aparecida Crispim et al., "Relationship between Food Intake and Sleep Pattern in Healthy Individuals," *Journal of Clinical Sleep Medicine* 7, no. 6 (2011): 659.
11. E. Donga et al., "A Single Night of Partial Sleep Deprivation Induces Insulin Resistance in Multiple Metabolic Pathways in Healthy Subjects," *Journal of Endocrinology Metabolism* 95, no. 6 (2010): 2963–68.
12. University of Chicago Medical Center, "Weekend Catch-Up Sleep Can Reduce Diabetes Risk Associated with Sleep Loss," ScienceDaily, January 18, 2016, https://www .sciencedaily.com/releases/2016/01/160118184342.htm.
13. S. M. Schmid et al., "A Single Night of Sleep Deprivation Increases Ghrelin Levels and Feelings of Hunger in Normal-Weight Healthy Men," *Journal of Sleep Research* 17, no. 3 (2008): 3313–14.
14. M. Dirlewanger et al., "Effects of Short-Term Carbohydrate or Fat Overfeeding on Energy Expenditure and Plasma Leptin Concentrations in Healthy Female Subjects," *International Journal of Obesity* 24, no. 11 (2000): 1413–18; M. Wabitsch et al., "Insulin

and Cortisol Promote Leptin Production in Cultured Human Fat Cells," *Diabetes* 45, no. 10 (January 1996): 1435–38.

15. W. A. Banks et al., "Triglycerides Induce Leptin Resistance at the Blood-Brain Barrier," *Diabetes* 53, no. 5 (2004): 1253–60.
16. E. A. Lawson et al., "Leptin Levels Are Associated with Decreased Depressive Symptoms in Women across the Weight Spectrum, Independent of Body Fat," *Clinical Endocrinology—Oxford* 76, no. 4 (2012): 520–25.
17. L. D. Baker et al., "Effects of Growth Hormone–Releasing Hormone on Cognitive Function in Adults with Mild Cognitive Impairment and Healthy Older Adults: Results of a Controlled Trial," *Archives of Neurology* 69, no. 11 (2012): 1420–29.
18. Helene Norrelund, "The Metabolic Role of Growth Hormone in Humans with Particular Reference to Fasting," *Growth Hormone & IGF Research* 15, no. 2 (2005): 95–122.
19. Intermountain Medical Center, "Routine Periodic Fasting Is Good for Your Health, and Your Heart, Study Suggests," ScienceDaily, May 20, 2011, https://www.sciencedaily.com/releases/2011/04/110403090259.htm.
20. Kukkonen-Harjula et al., "Haemodynamic and Hormonal Responses."
21. S. Debette et al., "Visceral Fat Is Associated with Lower Brain Volume in Healthy Middle-Aged Adults," *Annals of Neurology* 68, no. 2 (2010): 136–44.
22. E. S. Epel et al., "Stress and Body Shape: Stress-Induced Cortisol Secretion Is Consistently Greater among Women with Central Fat," *Psychosomatic Medicine* 62, no. 5 (2000): 623–32.
23. W. Turakitwanakan, C. Mekseepralard, and P. Busarakumtragul, "Effects of Mindfulness Meditation on Serum Cortisol of Medical Students," *Journal of the Medical Association of Thailand* 96, supplement 1 (2013): S90–95.
24. R. Berto, "The Role of Nature in Coping with Psycho-Physiological Stress: A Literature Review on Restorativeness," *Behavioral Sciences* 4, no. 4 (2014): 394–409.
25. T. Watanabe et al., "Green Odor and Depressive-like State in Rats: Toward an Evidence-Based Alternative Medicine?" *Behavioural Brain Research* 224, no. 2 (2011): 290–96.
26. C. D. Conrad, "Chronic Stress-Induced Hippocampal Vulnerability: The Glucocorticoid Vulnerability Hypothesis," *Reviews in the Neurosciences* 19, no. 6 (2008): 395–411.
27. J. J. Kulstad et al., "Effects of Chronic Glucocorticoid Administration on Insulin-Degrading Enzyme and Amyloid-Beta Peptide in the Aged Macaque," *Journal of Neuropathology & Experimental Neurology* 64, no. 2 (2005): 139–46.

GENIUS FOOD #9: WILD SALMON

1. Staubo, "Mediterranean Diet."

CHAPTER 10: THE VIRTUES OF STRESS (OR, HOW TO BECOME A MORE ROBUST ORGANISM)

1. Elsevier Health Sciences, "Prolonged Daily Sitting Linked to 3.8 Percent of All-Cause Deaths," EurekAlert!, March 26, 2016, https://www.eurekalert.org/pub_releases/2016-03/ehs-pds032316.php.
2. University of Utah Health Sciences, "Walking an Extra Two Minutes Each Hour May Offset Hazards of Sitting Too Long," ScienceDaily, April 30, 2015, https://www.sciencedaily.com/releases/2015/04/150430170715.htm.
3. University of Cambridge, "An Hour of Moderate Exercise a Day Enough to Counter

Health Risks from Prolonged Sitting," ScienceDaily, July 27, 2016, https://www.sciencedaily.com/releases/2016/07/160727194405.htm.

4. Kirk Erickson et al., "Exercise Training Increases Size of Hippocampus and Improves Memory," *Proceedings of the National Academy of Sciences* 108, no. 7 (2010): 3017–22.
5. Dena B. Dubal et al., "Life Extension Factor Klotho Enhances Cognition," *Cell Reports* 7, no. 4 (2014): 1065–76.
6. Keith G. Avin et al., "Skeletal Muscle as a Regulator of the Longevity Protein, Klotho," *Frontiers in Physiology* 5 (2014).
7. J. C. Smith et al., "Semantic Memory Functional MRI and Cognitive Function after Exercise Intervention in Mild Cognitive Impairment," *Journal of Alzheimer's Disease* 37, no. 1 (2013).
8. Jennifer Steiner et al., "Exercise Training Increases Mitochondrial Biogenesis in the Brain," *Journal of Applied Physiology* 111, no. 4 (2011): 1066–71.
9. "Fit Legs Equals Fit Brain, Study Suggests," BBC.com, November 10, 2015, http://www.bbc.com/news/health-34764693.
10. Mari-Carmen Gomez-Cabrera et al., "Oral Administration of Vitamin C Decreases Muscle Mitochondrial Biogenesis and Hampers Training-Induced Adaptations in Endurance Performance," *The American Journal of Clinical Nutrition* 87, no. 1 (2008): 142–49.
11. "Housing," Statistics Finland, May 15, 2017, http://www.stat.fi/tup/suoluk/suoluk_asuminen_en.html.
12. M. Goekint et al., "Influence of Citalopram and Environmental Temperature on Exercise-Induced Changes in BDNF," *Neuroscience Letters* 494, no. 2 (2011): 150–54.
13. Mark Maynard et al., "Ambient Temperature Influences the Neural Benefits of Exercise," *Behavioural Brain Research* 299 (2016): 27–31.
14. Simon Zhornitsky et al., "Prolactin in Multiple Sclerosis," *Multiple Sclerosis Journal* 19, no. 1 (2012): 15–23.
15. Laatikainen, "Response of Plasma Endorphins."
16. Wouter van Marken Lichtenbelt et al., "Healthy Excursions outside the Thermal Comfort Zone," *Building Research & Information* 45, no. 7 (2017): 1–9.
17. Denise de Ridder et al., "Always Gamble on an Empty Stomach: Hunger Is Associated with Advantageous Decision Making," *PLOS ONE* 9, no. 10 (2014): E111081.
18. M. Alirezaei et al., "Short-Term Fasting Induces Profound Neuronal Autophagy," *Autophagy* 6, no. 6 (2010): 702–10.
19. Megumi Hatori et al., "Time-Restricted Feeding without Reducing Caloric Intake Prevents Metabolic Diseases in Mice Fed a High-Fat Diet," *Cell Metabolism* 15, no. 6 (2012): 848–60.
20. F. B. Aksungar, A. E. Topkaya, and M. Akyildiz, "Interleukin-6, C-Reactive Protein and Biochemical Parameters during Prolonged Intermittent Fasting," *Annals of Nutrition and Metabolism* 51, no. 1 (2007): 88–95; J. B. Johnson et al., "Alternate Day Calorie Restriction Improves Clinical Findings and Reduces Markers of Oxidative Stress and Inflammation in Overweight Adults with Moderate Asthma," *Free Radical Biology & Medicine* 42, no. 5 (2007): 665–74.
21. Kauwe, "Acute Fasting."
22. Gary Wisby, "Krista Varady Weighs In on How to Drop Pounds," UIC Today, February 5, 2013, https://news.uic.edu/krista-varady-weighs-in-on-how-to-drop-pounds-fast.

23. Jan Moskaug et al., "Polyphenols and Glutathione Synthesis Regulation," *American Journal of Clinical Nutrition* 81, no. 1 (2005): 2775–835.
24. P. G. Paterson et al., "Sulfur Amino Acid Deficiency Depresses Brain Glutathione Concentration," *Nutritional Neuroscience* 4, no. 3 (2001): 213–22.
25. Caroline M. Tanner et al., "Rotenone, Paraquat, and Parkinson's Disease," *Environmental Health Perspectives* 119, no. 6 (2011): 866–72.
26. Claudiu-Ioan Bunea et al., "Carotenoids, Total Polyphenols and Antioxidant Activity of Grapes (*Vitis vinifera*) Cultivated in Organic and Conventional Systems," *Chemistry Central Journal* 6, no. 1 (2012): 66.
27. Vanderbilt University Medical Center, "Eating Cruciferous Vegetables May Improve Breast Cancer Survival," ScienceDaily, April 3, 2012, https://www.sciencedaily.com/releases/2012/04/120403153531.htm.
28. B. E. Townsend and R. W. Johnson, "Sulforaphane Reduces Lipopolysaccharide-Induced Proinflammatory Markers in Hippocampus and Liver but Does Not Improve Sickness Behavior," *Nutritional Neuroscience* 20, no. 3 (2017): 195–202.
29. K. Singh et al., "Sulforaphane Treatment of Autism Spectrum Disorder (ASD)," *Proceedings of the National Academy of Science USA* 111, no. 43 (2014): 15550–55.

GENIUS FOOD #10: ALMONDS

1. Z. Liu et al., "Prebiotic Effects of Almonds and Almond Skins on Intestinal Microbiota in Healthy Adult Humans," *Anaerobe* 26 (2014): 1–6.
2. A. Wu, Z. Ying, and F. Gomez-Pinilla, "The Interplay between Oxidative Stress and Brain-Derived Neurotrophic Factor Modulates the Outcome of a Saturated Fat Diet on Synaptic Plasticity and Cognition," *European Journal of Neuroscience* 19, no. 7 (2004): 1699–707.
3. A. J. Perkins et al., "Association of Antioxidants with Memory in a Multiethnic Elderly Sample Using the Third National Health and Nutrition Examination Survey," *American Journal of Epidemiology* 150, no. 1 (1999): 37–44.
4. R. Yaacoub et al., "Formation of Lipid Oxidation and Isomerization Products during Processing of Nuts and Sesame Seeds," *Journal of Agricultural and Food Chemistry* 56, no. 16 (2008): 7082–90.
5. A. Veronica Witte et al., "Effects of Resveratrol on Memory Performance, Hippocampal Functional Connectivity, and Glucose Metabolism in Healthy Older Adults," *Journal of Neuroscience* 34, no. 23 (2014): 7862–70.

CHAPTER 11: THE GENIUS PLAN

1. Tao Huang et al., "Genetic Susceptibility to Obesity, Weight-Loss Diets, and Improvement of Insulin Resistance and Beta-Cell Function: The POUNDS Lost Trial," American Diabetes Association—76th Scientific Sessions (2016).
2. Karina Fischer et al., "Cognitive Performance and Its Relationship with Postprandial Metabolic Changes after Ingestion of Different Macronutrients in the Morning," *British Journal of Nutrition* 85, no. 3 (2001): 393–405.
3. E. Fiedorowicz et al., "The Influence of μ-Opioid Receptor Agonist and Antagonist Peptides on Peripheral Blood Mononuclear Cells (PBMCs)," *Peptides* 32, no. 4 (2011): 707–12.
4. Anya Topiwala et al., "Moderate Alcohol Consumption as Risk Factor for Adverse Brain Outcomes and Cognitive Decline: Longitudinal Cohort Study," *BMJ* 357 (2017): j2353.

5. P. N. Prinz et al., "Effect of Alcohol on Sleep and Nighttime Plasma Growth Hormone and Cortisol Concentrations," *Journal of Clinical and Endocrinology and Metabolism* 51, no. 4 (1980): 759–64.
6. S. D. Pointer et al., "Dietary Carbohydrate Intake, Insulin Resistance and Gastro-oesophageal Reflux Disease: A Pilot Study in European- and African-American Obese Women," *Alimentary Pharmacology & Therapeutics* 44, no. 9 (2016): 976–88.
7. St-Onge, "Fiber and Saturated Fat."

CHAPTER 12: RECIPES AND SUPPLEMENTS

1. William Shankle et al., "CerefolinNAC Therapy of Hyperhomocysteinemia Delays Cortical and White Matter Atrophy in Alzheimer's Disease and Cerebrovascular Disease," *Journal of Alzheimer's Disease* 54, no. 3 (2016): 1073–84.
2. Ibid.

INDEX

ABOUT THE AUTHORS

Max Lugavere

Max Lugavere is a filmmaker, TV personality, and health and science journalist. He is the director of the film *Bread Head*, the first-ever documentary about dementia prevention through diet and lifestyle. Lugavere has contributed to Medscape, Vice, Fast Company, and the Daily Beast and has been featured on *NBC Nightly News*, *The Dr. Oz Show*, and *The Doctors* and in the *Wall Street Journal*. He is a sought-after speaker, invited to lecture at esteemed academic institutions such as the New York Academy of Sciences and Weill Cornell Medicine, and he has given keynotes at such events as the Biohacker Summit in Stockholm, Sweden. From 2005 to 2011, Lugavere was a journalist for Al Gore's Current TV. He lives in New York City and Los Angeles.

Paul Grewal, MD

Paul Grewal, MD, is an internal medicine physician and speaker who focuses on diet and lifestyle strategies for weight loss, metabolic health, and longevity. Having lost and kept off nearly one hundred pounds himself, he helps others find a sustainable, holistic, and enjoyable path to health, his greatest pride and passion. He earned a bachelor of arts in cellular and molecular neuroscience from Johns Hopkins University, studied medicine at Rutgers Medical School, and completed his residency at North Shore-Long Island Jewish Hospital. He is the founder of MyMD Medical Group, a private practice in New York City, and serves as a medical adviser to financial firms and health-care start-ups in the NYC area.